Freeze-Drying Technology in Foods

Freeze-Drying Technology in Foods

Editors

Valentina Prosapio
Estefania Lopez-Quiroga

MDPI • Basel • Beijing • Wuhan • Barcelona • Belgrade • Manchester • Tokyo • Cluj • Tianjin

Editors
Valentina Prosapio
University of Birmingham,
School of Chemical Engineering
UK

Estefania Lopez-Quiroga
University of Birmingham,
School of Chemical Engineering
UK

Editorial Office
MDPI
St. Alban-Anlage 66
4052 Basel, Switzerland

This is a reprint of articles from the Special Issue published online in the open access journal *Foods* (ISSN 2304-8158) (available at: https://www.mdpi.com/journal/foods/special_issues/Freeze_Drying_Technology).

For citation purposes, cite each article independently as indicated on the article page online and as indicated below:

LastName, A.A.; LastName, B.B.; LastName, C.C. Article Title. *Journal Name* **Year**, *Volume Number*, Page Range.

ISBN 978-3-0365-0068-3 (Hbk)
ISBN 978-3-0365-0069-0 (PDF)

Contents

About the Editors

Valentina Prosapio joined the University of Birmingham in 2016 as a Research Fellow in the School of Chemical Engineering. She works within the Microstructure Engineering Research Group with a focus on drying and rehydration. Her research involves the drying of foods using different techniques with the aim of preserving the properties of fresh products, improving the rehydration ability of dried products and decreasing the energy costs and environmental impact of the process. Her work also includes the encapsulation/incorporation of active compounds into polymeric carriers for controlled drug delivery.

Estefania Lopez-Quiroga is a Lecturer in the School of Chemical Engineering and the Centre for Doctoral Training in Formulation Engineering. Her research focuses on the development of digital manufacture processes and chains for formulated and/or structured products (e.g., foods, FMCG, pharma), by integrating novel modeling and computational tools with (bio)process engineering methods.

Editorial

Freeze-Drying Technology in Foods

Valentina Prosapio * and Estefania Lopez-Quiroga

School of Chemical and Engineering, University of Birmingham, Birmingham B15 2TT, UK;
E.Lopez-Quiroga@bham.ac.uk
* Correspondence: v.prosapio@bham.ac.uk

Received: 21 May 2020; Accepted: 8 July 2020; Published: 13 July 2020

Keywords: freeze-drying; process design; rehydration; modelling; microstructure; food quality; encapsulation; processes combination

---

Freeze-drying (or lyophilisation) is a drying method, largely employed in the food industry. It consists in the freezing of the product, followed by sublimation of the ice at reduced pressure. Among drying processes, freeze-drying is considered to be the gentlest one, as it causes negligible damage to the product microstructure, allowing fast rehydration rates and high rehydration capacity, and good preservation of the physical chemical properties.

Despite being a common technique, lot of research on freeze-drying is still ongoing to optimise the process conditions according to specific applications, to improve the product characteristics by applying pre-treatments, and to reduce the energy costs and processing time.

The Special Issue "Freeze-Drying Technology in Foods" focuses on the application of freeze-drying on food and nutraceutical fields and groups four original studies and one review.

The study carried out by Silva-Espinoza et al. [1] focused on the optimisation of the freeze-drying operating conditions to better preserve the physical-chemical properties of orange puree. The authors investigated the effect of freezing rate (conventional and blast freezer), working pressure (5–100 Pa) and shelf temperature (30–50 °C) on quality parameters such as colour, porosity, mechanical properties, water content, vitamin C, total phenols, β-carotene and antioxidant activity. Colour analyses showed that colour was better preserved when the highest operating pressure, highest temperature, and fast freezing conditions were used. Freeze-dried purees showed a high degree of porosity, but sample porosities obtained at different conditions were not statistically different. Mechanical analysis showed that the higher mechanical resistance of the sample is achieved working at the lowest pressure and the highest temperature. In addition, samples with high mechanical resistance to fracture showed the smallest moisture content. Lower degradation of the nutrients was then observed at a higher temperature due to the faster completion of the drying process. The authors concluded that the optimal freeze-drying conditions to maximise the quality of freeze-dried orange puree are low pressure and high temperature.

Munzenmayer et al. [2] studied the effect of applying CO_2 laser microperforations to blueberry skin prior freeze-drying. The results showed that the primary drying time was significantly reduced from 17h for non-treated berries to 13h when nine microperforations per berry were applied, with minimal effect on the fruit appearance. At the same time, the fruit quality was also significantly improved, as the percentage of non-busted blueberries at the end of the process increased from 47% to 86%. In fact, the authors showed that the microholes work as pathways for the escape of vapour from the sublimating front through the weakened mass transfer resistance of the blueberry skin, relieving the pressure development underneath, eventually avoiding the fruit bust and improving the quality of the resulting product with a reduced processing time.

Prosapio et al. [3] applied freeze-drying to gellan gun gels loaded with vitamin B_2 and investigated the effect of the gel pH on drying and release kinetics. They observed that acidified gellan gum gels at pH 2.5 showed the fastest drying rate, whereas gels at pH 4 showed the slowest rate. For natural pH samples (5.2), the Page model provided the most accurate description of freeze-drying kinetics, whereas the Wang and Singh model predicted more accurately the kinetics at pH 4 and 2.5. The authors investigated also the effect of the gel pH on the vitamin release mechanism using the Korsmeyer–Peppas model. Freeze-dried gels at pH 4 completed the vitamin release in about 9.5 h; gels at natural pH in 6h, while samples at pH 2.5 in 3 h. These differences were ascribed to the different gel microstructure. Freeze-dried gellan gum gels at pH 4 exhibited an aggregated and rigid structure that can impede mass transfer within the gel, increasing the time needed to release the vitamin completely from the substrate. Samples at pH 2.5 presented a low aggregated and weak structure that lead to breakage during the release experiments making the vitamin delivery faster. Natural pH gels showed an intermediate behaviour due to an intermediate level of aggregation. The Korsmeyer–Peppas model was used to analyse experimental release curves, revealing that samples at pH 5.2 display a typical Fickian behaviour, while acidified samples at pH 4 have combined both Fickian and relaxation mechanisms.

He et al. [4] investigated the emulsion properties of Aquafaba (AQ), a viscous by-product solution produced during cooking chickpea or other legumes in water. They carried out a screening on the different chickpea cultivars grown in Canada (CDC Leader, CDC Orion, CDC Luna, CDC Consul and Amit) and studied the impact of chickpea seed physicochemical properties (Seed coat incidence, Seed dimensions, surface area per unit mass of seed and seed coat weight per surface area) and hydration kinetics on the properties of AQ-based emulsions. The authors showed that the type of cultivar has a significant effect on the emulsion capacity and stability, being these values the highest when the CDC Leader cultivar was employed. In contrast, a correlation between the composition of the chickpea seed (carbohydrates, proteins, fat) and the emulsion properties was not observed.

Bhatta et al. [5] conducted a review on the application of freeze-drying to plant-based foods which presents the most recent research publications on the subject and also includes original research on the topic by the authors. This work recalls the principle of freeze-drying and the main features of plant-based foods, highlighting their advantages and challenges upon lyophilization, and offers a thorough overview of the applications of freeze-dried fruits, vegetables and speciality foods (coffee, tea, spices). It follows a discussion of the most common studied materials and the effect of the process conditions (shelf temperature, pressure, processing time, sample size/cut) on product quality (shrinkage, porosity, colour, rehydration, nutrients retention, moisture content, thermal properties, morphology, texture, powder flowability). Finally, and with the aim of revealing potential improvements to product quality and process optimization in plant-based foods, the authors carried out a critical analysis of the most common pre-treatments: (i) chemical, i.e., immersion of the product in alkaline or acid solutions of oleate esters prior to drying; (ii) mechanical, consisting in the peeling, abrasion of the surface, puncturing the skin, or cutting the fruit in various shapes; (iii) thermal, involving blanching or steaming; (iv) freezing pre-treatments, including individual quick freezing, freeze-granulation, foaming, and the application of infrared energy, ultrasounds and microwaves.

In summary, the five papers published in this Special Issue highlighted the main challenges in the ongoing research on freeze-drying and identified strategies to make this method more convenient and to better preserve the product quality with the aim of meeting industry needs and consumer expectations.

Author Contributions: Authors V.P. and E.L.-Q. contributed equally to this editorial article. All authors have read and agreed to the published version of the manuscript.

Funding: E. Lopez-Quiroga acknowledges financial support from the EPSRC Centre for Doctoral Training in Formulation Engineering (grant number EP/S023070/1).

References

1. Silva-Espinoza, M.A.; Ayed, C.; Foster, T.; Camacho, M.D.M.; Martínez-Navarrete, N. The Impact of Freeze-Drying Conditions on the Physico-Chemical Properties and Bioactive Compounds of a Freeze-Dried Orange Puree. *Foods* **2020**, *9*, 32. [CrossRef] [PubMed]
2. Munzenmayer, P.; Ulloa, J.; Pinto, M.; Ramírez, C.; Valencia, P.; Simpson, R.; Almonacid, S. Freeze-Drying of Blueberries: Effects of Carbon Dioxide (CO2) Laser Perforation as Skin Pretreatment to Improve Mass Transfer, Primary Drying Time, and Quality. *Foods* **2020**, *9*, 211. [CrossRef] [PubMed]
3. Prosapio, V.; Norton, I.T.; Lopez-Quiroga, E. Freeze-Dried Gellan Gum Gels as Vitamin Delivery Systems: Modelling the Effect of pH on Drying Kinetics and Vitamin Release Mechanisms. *Foods* **2020**, *9*, 329. [CrossRef] [PubMed]
4. He, Y.; Shim, Y.Y.; Mustafa, R.; Meda, V.; Reaney, M.J. Chickpea Cultivar Selection to Produce Aquafaba with Superior Emulsion Properties. *Foods* **2019**, *8*, 685. [CrossRef]
5. Bhatta, S.; Stevanovic, T.; Ratti, C. Freeze-Drying of Plant-Based Foods. *Foods* **2020**, *9*, 87. [CrossRef] [PubMed]

Review

Freeze-Drying of Plant-Based Foods

Sagar Bhatta [1], Tatjana Stevanovic Janezic [1] and Cristina Ratti [2,*]

[1]   Département Sciences du Bois et de la Forêt, Institute of Nutrition and Functional Foods, Université Laval, Quebec City, QC G1V 0A6, Canada; sagar.bhatta.1@ulaval.ca (S.B.); Tatjana.Stevanovic@sbf.ulaval.ca (T.S.J.)
[2]   Département des Sols et de Génie Agroalimentaire, Institute of Nutrition and Functional Foods, Université Laval, Quebec City, QC G1V 0A6, Canada
*   Correspondence: Cristina.Ratti@fsaa.ulaval.ca; Tel.: +1-418-656-2131; Fax: +1-418-656-3723

Received: 29 November 2019; Accepted: 9 January 2020; Published: 13 January 2020

Abstract: Vacuum freeze-drying of biological materials is one of the best methods of water removal, with final products of highest quality. The solid state of water during freeze-drying protects the primary structure and the shape of the products with minimal volume reduction. In addition, the lower temperatures in the process allow maximal nutrient and bioactive compound retention. This technique has been successfully applied to diverse biological materials, such as meats, coffee, juices, dairy products, cells, and bacteria, and is standard practice for penicillin, hormones, blood plasma, vitamin preparations, etc. Despite its many advantages, having four to ten times more energy requirements than regular hot air drying, freeze-drying has always been recognized as the most expensive process for manufacturing a dehydrated product. The application of the freeze-drying process to plant-based foods has been traditionally dedicated to the production of space shuttle goods, military or extreme-sport foodstuffs, and specialty foods such as coffee or spices. Recently, the market for 'natural' and 'organic' products is, however, strongly growing as well as the consumer's demand for foods with minimal processing and high quality. From this perspective, the market for freeze-dried plant-based foods is not only increasing but also diversifying. Freeze-dried fruits and vegetables chunks, pieces, or slices are nowadays majorly used in a wide range of food products such as confectionaries, morning cereals, soups, bakeries, meal boxes, etc. Instant drinks are prepared out of freeze-dried tea, coffee, or even from maple syrup enriched with polyphenol concentrated extracts from trees. The possibilities are endless. In this review, the application of freeze-drying to transform plant-based foods was analyzed, based on the recent research publications on the subject and personal unpublished data. The review is structured around the following related topics: latest applications of freeze-drying to plant-based foods, specific technological problems that could be found when freeze-drying such products (i.e., presence of cuticle; high sugar or lipid concentration), pretreatments and intensification technologies employed in freeze-drying of plant-based foods, and quality issues of these freeze-dried products.

Keywords: freeze-drying; lyophilization; plant-based foods; fruits; vegetables

---

1. Introduction

Plant-based foods, including fruits, vegetables, seeds, beans, spices, etc., are important components of a healthy diet, and their sufficient regular consumption could help to prevent certain major diseases such as cancer and cardiovascular diseases, etc. According to the combined report of World Health Organization and Food and Agriculture Organization, it was recommended that a daily minimum consumption of 400 g of fruits and vegetables may help to minimize the occurrence of chronic diseases along with the mitigation of micronutrient deficiencies [1]. Fresh plant-based foods may not available all year round for consumption and the long-term storage of fresh foods could be challenging due to high water content, unavailability of cold-storage facilities (particularly in underdeveloped and developing countries), and possibility of nutritional deterioration. Consequently, drying of such foods may allow their long-term consumption and eases handling, transportation, and storage.

Freeze drying (FD), also known as lyophilization, is a well-known technique for the production of high quality food powders and solids [2,3]. It is a preferred method for drying foods containing compounds that are thermally sensitive and prone to oxidation since it operates at low temperatures and under high vacuum. Application of FD to various plant-based foods, such as apple, guava, strawberry, blackberry, pumpkin, tomato, asparagus, coffee, tea, garlic, ginger, maple syrup, etc., has already been reported in literature [4–15].

Drying of food results in food quality changes [3]. The quality of foods can be divided into three sections: physical, chemical, and nutritional. Major qualities of foods that can be affected during drying processes are color, odor, texture, rehydration property, bulk properties, flow property, water activity, and retention of nutrients and volatile compounds [16]. Regarding to nutritional qualities, oxygen, high temperature, and cell damage are usual enemies of bioactive compound retention during processing. The stability of the valuable compounds of plant-based foods can therefore be affected during dehydration. Phenolic compounds could be susceptible to enzymatic degradation due to the polyphenol oxidase activity [17]. In addition, carotenoids have a highly unsaturated nature, making them susceptible to degradation by oxidation and thermal processes. Oxidation is the major cause of carotenoids degradation and can be generally considered autocatalytic, beginning only after an induction period in which radicals are built up and antioxidants are depleted [18]. Also, the loss of vitamin C and carotenoids is affected particularly by the temperature and the moisture content during drying processes [19]. Vitamin C is usually selected as a reference index of the nutrient quality due to its labile nature compared to other nutrients in foods [20], and thus, if ascorbic acid is well retained, other nutrients would be as well. A few interesting reviews on the impact of drying methods and operating conditions on functional quality retention can be found in the literature [16,21].

Freeze-drying method by lack of liquid water, oxygen-free environment (if operated under vacuum), and low operating temperatures is thus the best choice to dehydrate fruits and vegetables in order to keep an optimized biocompound content in the final products. Despite the long drying time and expensive process, freeze-drying is widely used to produce high-value food products due to maximal retention of food quality when compared to other drying techniques. FD is considered as the standard or reference drying method in most research studies. Lately, various process intensification approaches have been implemented in order to overcome the challenges facing by FD methods, such as either pretreatment of the sample or the use of innovative technologies including infrared, microwave, and ultrasound energy with freeze-drying.

The aim of current review was to highlight the application of FD of plant-based foods, to point out some particular technological challenges, and describe process intensification in FD of plant-based foods to improve the quality of freeze-dried foods or to accelerate the process.

2. Methodology

The following databases were used for a bibliographic research: Web of Science (2000–2019) and Google Scholar (2000–2019). Authors have presented some of their original works (experimental data and images) to support conclusions.

3. FD Principle

Water exists in three different states: solid, liquid, or gas (vapor). Figure 1 presents the phase diagram of water (pressure versus temperature), where the curve lines show the passage from solid to vapor (sublimation), from liquid to vapor (evaporation), or from solid to liquid (fusion). Point T in Figure 1 represents the triple point of water (at 0.01 °C and 0.612 kPa) where the three phases (liquid, vapor, solid) coexist, and point C is the critical point of water (374 °C and 22060 kPa). Freeze-drying makes use of the sublimation phenomenon (at temperatures lower than 0.01 °C, and water vapor pressures below 0.612 kPa). In Figure 1, a product to be freeze-dried will follow the path from A to point B (i.e., the product should be first frozen by decreasing its temperature, then the water vapor pressure should be lowered below the pressure corresponding to the triple point, and finally some heat should be supplied to help the ice to convert into vapor by sublimation).

During the FD process, the removal of solid-state water (ice) occurs in three steps: (a) freezing, where the sample should be completely frozen; (b) primary drying, when ice is sublimated, usually at sub-atmospheric pressure; and (c) secondary drying, when the remaining unfrozen/bound water is desorbed from the drier food matrix.

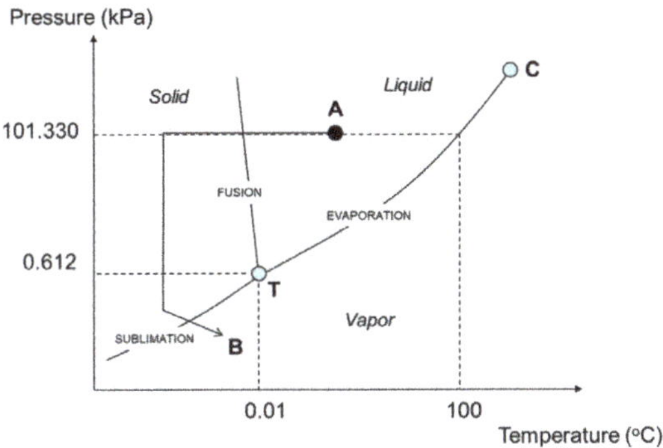

Figure 1. Phase diagram of water (T: triple point of water, C: critical point of water). "A" represents the starting point prior to freeze-drying (atmospheric pressure and ambient temperature), while "B", the desired final conditions during sublimation (below the triple point T).

Freezing is the first separation step in the FD process, which solidifies the food materials. The rate of freezing is important for the formation and size of ice crystals—slow rate of freezing forming bigger ice crystals and vice versa. Accordingly, the size of crystals affects the rate of drying, wherein large ice crystals are easier to sublimate and hence increase the rate of primary drying [22].

In primary drying, a vacuum is applied and the shelf temperature is increased to start the sublimation, such that the product temperature is 2–3 °C below the collapse temperature T_c [23,24]. Collapse temperature is the temperature above which the product has the risk of losing macroscopic structure during the FD process [24]. T_c could be determined with a freeze-drying microscope, but also may be estimated from the glass transition temperature (Tg). It should be noted that Tc could be 2 °C to 20 °C higher than Tg, depending mainly on sample composition [24,25]. However, very conservative predictions of the collapse temperature may only result in a much longer freeze-drying process, thus it can only be used in critical cases when the sample is difficult to freeze-dry. Figure 2 depicts the typical temperature profile of a product during each step of the freeze-drying process, where it can be observed that during primary drying the product temperature should be below the collapse temperature (represented as a dotted line T_1 in Figure 2).

Secondary drying starts when sublimation is still in place, being a slow part of the freeze-drying process, which may take at least 30% longer to complete than the end of sublimation. This last step could be performed at an elevated shelf temperature to more efficiently remove the remaining unfrozen or bound water by desorption, but lower than the glass transition temperature of dry solids (represented as dotted line T_2 in Figure 2). However, it is challenging to identify the endpoint of primary drying or the beginning of secondary drying phase. If temperature is increased before all the ice is sublimated (endpoint of primary drying phase), it could collapse the product and hence, affect the final quality. Patel et al. [23] have suggested some techniques to determine the endpoint of the primary drying such as Pirani pressure gauge, dew point monitor, tunable diode laser absorption spectroscopy (TDLAS), gas plasma spectroscopy, thermocouple (TC), and condenser pressure. Among these techniques, Pirani, dew point, TDLAS, and TC were found to be effective for determining the endpoint of the primary drying phase.

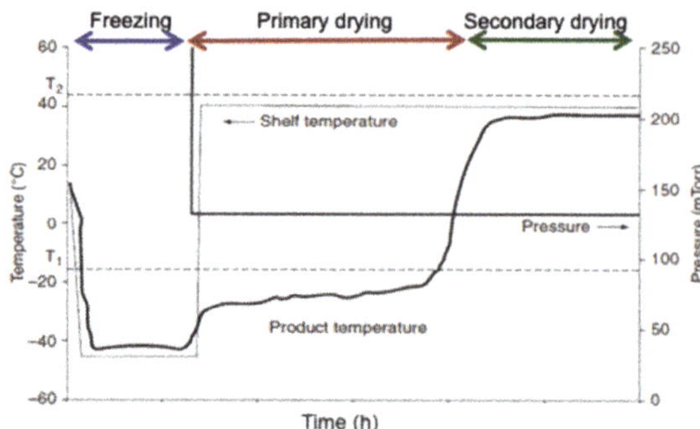

Figure 2. Temperature profile of product during freeze-drying process, where T_1 (dotted line) is the collapse temperature and T_2 (dotted line) is the glass transition temperature of dry solids (adapted from [22]).

4. Characteristics of Plant-Based Foods, Their Advantages, and Challenges upon Freeze-Drying

Plant-based foods are derived from vegetables, grains, nuts, seeds, legumes, and fruits [26]. Two types of plant-based foods are used in freeze-drying applications: solids and homogeneous solutions/suspensions such as juices or purees.

Solid plant foods present intrinsic characteristics in terms of structure, anatomy, and composition, which may pose challenges on one hand, but otherwise occasionally help to ease the freeze-drying operation. To start, solid plant-based foods are mainly cellular solids. Gibson [27] reported that apples and potatoes are examples of a simple cellular tissue: parenchyma with thin-walled, polyhedral cells resembling an engineering closed-cell foam, as shown in Figure 3 for potato [28]. Unlike solutions and colloidal systems, cellular solids present stronger mechanical attributes related to the properties of the cell wall material and to the cell geometry—cellular materials allowing the simultaneous optimization of stiffness, strength, and overall weight in a given application [29]. Cellulose and noncellulosic (hemi-celluloses and pectic) polysaccharides are the main polymers forming the cell wall of plant-based foods. Cellulose is the single most abundant polysaccharide component of vegetable cell walls, presenting areas of crystallinity imparting a considerable tensile strength close to 1 GPa, and with a Young's modulus roughly 130 GPa [27,30]. Also, the mechanical response of cellular materials is enhanced by their arrangement and local geometry [31]. In this sense, if the freezing step is properly done at adequate low temperatures (without ice crystals destroying/weakening the cell walls), cellular materials are better prepared to stand during freeze-drying. It can be said that when freeze-drying solid cellular foods, mechanical properties and structural strength may play a more important role in keeping product integrity than glass transition temperature in order to avoid collapse during primary/secondary drying and storage of freeze-dried foods.

Most plants present an epidermis in their outer parts serving against water loss, regulating gas exchange, and secreting metabolic compounds to protect internal tissues against diseases and acting as a natural insect repellent as well. Figure 4 shows the cross section of the epidermal cuticle of *Vaccinium angustifolium* (lowbush) blueberries (magnified 250 times) [32]. This epidermis formed by a lipidic hydrophobic cuticle layer [33] constitutes an interface between the internal cells and the external environment, acting as a moisture barrier during growing, which enormously affects the water diffusion during subsequent processing, decreasing significantly the rate of freeze-drying when the whole plant-based material is dried (i.e., the case of berries/grapes). The outer surface of the cuticle is covered by epicuticular waxes (a lipid-soluble fraction) and consists of complex mixtures of long-chain aliphatic and cyclic components, including primary alcohols (C26, C28, C30), hydrocarbons (C29, C31), esters, fatty acids, and triterpenoids [34,35]. Intracuticular waxes are embedded in the cutin polymer matrix itself (a lipid-insoluble fraction), though little information is available on its composition [36]. This external waxy layer makes freeze-drying of whole fruits/vegetables challenging since vapor generated by ice sublimation during the primary step is trapped inside the product, increasing its pressure and thus, melting the ice. Finally, after a continuous pressure build-up, the product cracks or explodes inside the freeze-dryer, depending on the vacuum level. The quality of such freeze-dried product is therefore unacceptable, and thus, pretreatments are required to overcome this problem (please refer to Section 6).

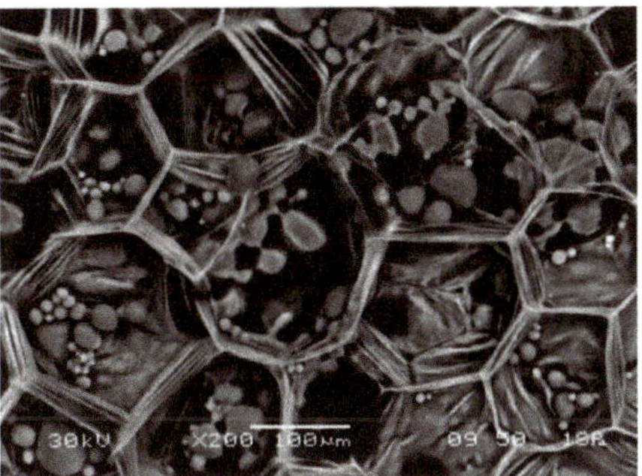

Figure 3. Cellular structure of potato.

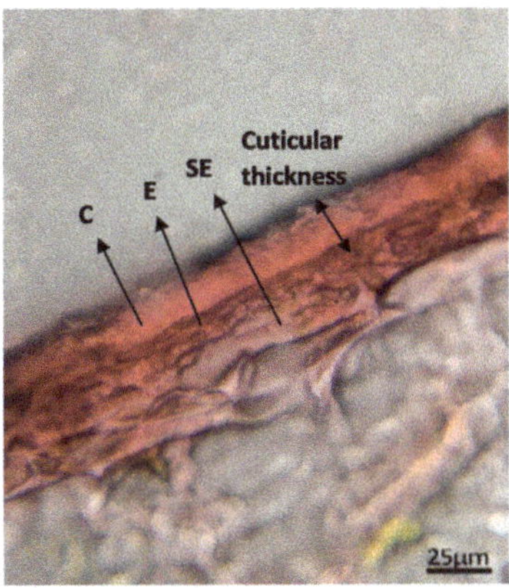

Figure 4. Optical microscope photo of a blueberry epidermis zone, where *C* is the proper cuticle, *E* is the epidermis, and *SE* is the sub epidermis.

For plant-based liquid solutions, sample composition may affect freeze-drying operation depending on one hand by the type of compound and its impact on overall glass transition temperature, but also on the total concentration of solids.

In a glass state, the viscosity of the matrix is high and the molecular movement is very limited. Glass transition occurs when a glassy matrix changes to a rubbery state, which is a more mobile amorphous structure. As explained previously, collapse temperature (related to the glass transition temperature Tg) represents the temperature above which the matrix loses its structure and the quality decreases, obviously related to the decrease in viscosity happening during glass transition. Therefore, when temperature during a process increases over the Tg of a product, the deterioration risk of many of its physical properties rises as well. Table 1 shows an example of the carbohydrate composition of apple and pear [37] together with literature values of the glass transition temperatures of pure sucrose, glucose, fructose, and D-sorbitol. As can be seen, pear juice has higher mass fractions of glucose and sorbitol than apple juice, and lower fractions of sucrose and fructose. Glass transition temperature of a multicomponent mixture could be roughly estimated using the following equation:

$$Tg = \sum_{i=1}^{n} x_i Tg_i \tag{1}$$

where x_i and Tg_i are the mass fraction and individual glass transition temperature of each component, respectively. Water is a plasticizer, having a low glass transition temperature, and some authors have indicated a value of −137 °C [38]. Thus, while drying takes place, the glass transition temperature of a product increases as water content is reduced, as indicated in Equation (1). Equation (1) usually underestimates experimental glass transition values [25], however, it can be used in this manuscript to predict the effect of composition for comparison purposes. From this equation, and using the mass fractions of Table 1, the predictions of glass transition temperature for dry pear juice (≈ 60.9 °C) are approximately 16 degrees lower than for apple juice (≈ 77.1 °C), probably due to its lower mass fraction of fructose and higher mass fraction of sorbitol. Experimental values for dried apple and pear showed a similar difference in glass transition temperatures values of an average of 10 degrees [39], thus validating the general conclusions obtained from Equation (1).

Table 1. Juice carbohydrate composition and glass transition temperatures.

Compound	Composition (% Total Carbohydrates)		Tg (°C) [40]
	Apple Juice (Gala Royal)	**Pear Juice (Bartlett)**	
Sucrose	11.7	4.7	52
Glucose	18.6	29.7	31
Fructose	65.4	49.6	100
D-Sorbitol	4.3	16.0	−2

From the previous discussion, it can be said that pear juice would have less thermal stability than apple juice upon freeze-drying at similar operating conditions, provoking final freeze-dried pear juice with lower quality (i.e., darker, stickier, lower rehydration, etc.). These predictive results have been corroborated by experimental freeze-drying data [37]. In these cases, the product with lower glass transition has to be freeze-dried at lower shelf temperatures and under higher vacuum, making the process longer and increasing costs. This example aimed to illustrate the utmost importance that composition, and its influence in glass transition, has for freeze-drying of liquid plant-based foods, such as juices. As indicated in [41], the collapse temperature of pure orange juice is relatively low, the dry juice collapsing at 52 °C. This collapse temperature is very close to that of sucrose (55 °C), due to the higher sucrose content of this juice (more than 50% of the sugars). In the same study, it was shown that addition of macromolecules increases the collapse temperature of freeze-dried orange juice, thus providing better thermal stability.

In cellular solid foods, collapse during drying takes place when the natural turgidity of product is lost and cannot be restored. The impact of glass transition in this case is less important since structure plays a major role in understanding the collapse phenomena. In a freeze-drying study of potato, celery, and apple at temperatures below, near, and above their Tg, [42] pointed out that differences in plant tissues (structure, composition) may contribute together with glass transition to prevent collapse.

In particular for liquids (solutions, emulsions, suspensions), the matrix of the product to be freeze-dried provides for 'body', mechanical strength, and an attractive appearance [43]. Concentration of simple constituents of this matrix could thus have a significant impact on the freeze-drying operation since high levels of sugars or lipids may convert to low quality final freeze-dried products. For instance, when the sugar content is too high (i.e., concentrated orange juice, maple syrup), freezing temperatures should be set up at lower optimized levels for a successful freezing step. However, even if the temperature of the freezing step has been reduced, sugars might migrate to the surface of the product during freezing, building up a barrier for water diffusion that will affect the further primary drying step, having a similar role as the epidermis of berries mentioned earlier. Thus, as water has difficulty in escaping from the matrix, pressure builds up and the surface may crack, which is a positive solution for letting water to escape, but ice might melt, with goods exploding inside the freeze-dryer, producing final products with undesirable characteristics. Pretreatments should be used in such cases, as explained in Section 6.

Spray-drying is a continuous process considered industrially viable to produce powders out of fluids due to its lower costs compared to freeze-drying. However, when dealing with plant-based foods prone to collapse or containing valuable oxidative compounds, spray-drying at high temperatures and using enormous amounts of air is not as effective as a freeze-drying. Expensive complicated formulations are required to be added to solutions to be spray-dried in order to avoid quality problems related to oxidation, or important yield losses due to collapse and stickiness.

5. Application of FD to Plant-Based Foods

5.1. Fruits

Table 2 presents the wide range of freeze-dried fruits that have been reported in literature, including strawberry, blackberry, guava, pineapple, etc. [10,12,14,44–48]. Fresh fruits containing high moisture levels are difficult to dehydrate by classical drying techniques due to significant damage to their physical attributes, mainly collapse and severe bleeding due to the rupture of skin. Shishehgarha et al. [14] studied the various quality parameters (drying kinetics, color, and volume variation) of sliced and whole strawberries under different FD shelf temperatures (30, 40, 50, 60, and 70 °C). The level of shrinkage was found to be independent of FD shelf temperature, where whole and sliced strawberries had an average volume reduction of 8% and 2%, respectively. On the other hand, the risk of collapse was found to increase exponentially once the FD shelf temperature surpassed the glass transition temperature of dried strawberries [14]. Shrinkage and collapse of fruits are often encountered during drying, mainly by air-drying techniques. Hawlader et al. [12] compared quality of guava obtained by heat pump dryers (RH = 10%, v = 0.7 m/s, T = 45 °C, 8 h), vacuum (vacuum pressure = 15,000 Pa, T = 45 °C, 8 h), and freeze-dryers (freezing at −20 °C for 24 h, and less than 613.2 Pa vacuum pressure, 10 °C shelf temperature, 24 h for freeze-drying). Porosity, color, rehydration, and vitamin C retention of guava produced by FD were better, resulting in FD guava being the most desirable powder compared to that produced by vacuum and heat pump dryers. FD was able to retain 63% vitamin C, whereas heat pump dryer retained only 25%.

It is often challenging to dehydrate fruits with waxy impermeable skin, as described previously, such as seabuckthorn berries (*Hippophae rhamnoides* L.), which are delicate fruits containing natural antioxidants, ascorbic acids, carotenoids, and flavonoids. Araya-Farias et al. [44] investigated the effect of hot air drying (HAD) at 1 m/s and 50 °C or 60 °C, and FD (4 Pa of vacuum, 20 °C or 50 °C shelf temperature) to obtain dried seabuckthorn pulp (shown in Table 2). They found that FD retained 93% more carotenoids, 34% more vitamin C, and 11% more phenolics than HAD. Increased drying temperature resulted in increase of drying kinetics. However, no significant impact of drying temperature on overall nutritional retention was noticed for FD or HAD samples, although a slight decrease was found for particular compounds such as vitamin E for both drying methods while increasing the operation temperature. Between two drying processes, it was reported that drying kinetics were surprisingly faster in FD.

5.2. Vegetables

Vegetables are good source of essential nutrition for human diet. Dehydration of vegetables is usually done for their long-term consumption. Studies on FD of some vegetables, including asparagus, carrot, pumpkin, and tomato, are presented in Table 2. Nindo et al. [13] studied the drying of sliced asparagus using FD and other drying methods including tray dryer, refractance window dryer, and spouted bed dryer. The highest amount of ascorbic acid was retained when samples were dehydrated by FD and refractive window drying.

Freeze-dried pumpkin has numerous applications in manufacturing formulated foods such as soups, noodles, breads, and cakes. Several authors have studied the FD of pumpkin to characterize its nutritional and physicochemical properties for above-mentioned food applications [6,49,50]. Guiné et al. [49] reported the decrease in moisture content from 90% to 8% in freeze-dried pumpkin, but FD induced a softening of the pumpkin, as hardness of pumpkin decreased from 19.37 N (fresh) to 1.59 N (dried/rehydrated) when the textural properties were analyzed. Also, the change in color was important (total change in color, $\Delta E = 12$). Ciurzyńska et al. [6] studied the effect of different pretreatment methods (blanching and osmotic dehydration) on the properties of freeze-dried pumpkin. The long duration of osmotic dehydration caused a decrease in water content of pumpkin and on water activity of final freeze-dried samples.

Gümüşay et al. [11] investigated the effects of sun, oven, vacuum-oven, and freeze-drying on the phenolic amount, antioxidant capacity, and ascorbic acid content of tomatoes. Freeze dried tomatoes had about twofold higher phenolic content than that of other drying methods (654.60 vs. 314.27–355.79 mg gallic acid equivalent/100 g dm). Unlike FD, processes using high drying temperatures may cause activation of oxidative enzymes, resulting in the loss of phenolic compounds. Enzymes such as peroxidative and hydrolytic may also have been liberated due to the disruption of tomato structure at high drying temperature [51]. The retention of ascorbic acid content (65.47 vs. 4.14–24.39 mg/100 g dm) and antioxidant capacity (1699.59 vs. 873.32–1148.86 mg trolox/100 g dm) was highest for freeze-dried tomatoes. Rajkumar et al. [52] reported high rehydration ratio and aroma retention in freeze-dried carrots. In terms of shrinkage, carrots exposed to FD had lower shrinkage rate (20.83%) than HAD (35.53%). Leafy vegetables such as spinach are also reported in the literature to be dehydrated by FD. An-Erl King et al. [53] found that the freeze-dried spinach had high porosity and surface area (263.6-296.8 m^2/g). The chlorophyll content of freeze-dried spinach decreased with an increase in storage temperature and storage time.

Table 2. Some examples of freeze-dried fruits and vegetables.

Food [Reference]	Sample Preparation	FD Conditions	Key Quality Studied
Acai [54]	n/a	n/a	Antioxidant activity
Asparagus [13]	2–4 mm slices	T (shelf) = 20 °C T (condenser) = –64 °C Pressure = 3.3 kPa (3300 Pa) Time= 18–24 h	Rehydration; color; antioxidant; ascorbic acid
Blackberries [10]	Juice with carrier agents	T (not reported if shelf or condenser) = –84 °C Pressure = 0.0004 Pa Time = 48 h	Moisture; Thermal property; density; morphology; antiradical activity
Carrot [55]	3–4 mm slices	T (shelf) = 30 °C T (condenser) = –60 °C Pressure = 6 Pa Time = n/a	Moisture content; carotenoid content; lycopene content
Chinese gooseberry [48]	3 mm in height and 4 mm in diameter	2-step FD protocol (T shelf, time) = –20 °C for 20 h and +20 °C for 5 h Pressure = 10 Pa	Moisture content; Thermal properties; Sorption isotherm
Date [47]	Date pulp with carrier agents	T (not reported if shelf or condenser) = –40 °C Pressure = 42 Pa Time = 72 h	Moisture; powder flowability; morphology; microstructure
Guava and papaya [12]	1 cm cubes	T (shelf) = 10 °C Pressure = less than 613.2 Pa Time = 24 h	Color; porosity; rehydration; texture; Vitamin C

Table 2. *Cont.*

Food [Reference]	Sample Preparation	FD Conditions	Key Quality Studied
Green Pepper [49]	2 cm × 2 cm	T (not reported if shelf or condenser) = –47 to –50 °C Pressure = 0.666 Pa Time = 38 h	Texture; color
Pumpkin [49]	2 cm × 2 cm	T (not reported if shelf or condenser) = –47 to –50 °C Pressure = 0.666 Pa Time = 38 h	Texture; color
Pumpkin [6]	10 mm cubes	T (shelf) = 10 °C Pressure = 63 Pa Time = 24 h	Moisture content; water activity; color
Seabuckthorn berries [44]	Pulp and seeds	T (shelf) = 20 or 50 °C Pressure = 4 Pa Time = 24 h	Drying kinetics; nutritional composition
Seabuckthorn berries/leaves/seeds [56]	Crushed	FD Process conditions = n/a Time = 48 h	Moisture; water and oil absorption; color; structure; antiradical activity
Strawberries [14]	Sliced or whole fruits	T (shelf) = 30–70 °C Pressure = 4 Pa Time = 12, 24, or 48 h	Color;volume;collapse
Strawberries [46]	Half-cut or Sliced	T (shelf) = 55 °C Pressure = 4 Pa Time = 28 h	Rehydration; color; firmness
Tomatoes/Ginger [11]	Sliced	T (not reported if shelf or condenser) = –50 °C Pressure = 0.001330 Pa Time = 24 h	Total phenolic; ascorbic acid; antioxidant capacity
Tropical fruits (pineapple, Barbados cherry, guava, papaya, and mango) [57]	125 mm in diameter and 5 mm in height	T (not clear which temperature it is) = –30 °C Pressure = 0.001330 Pa Time = 12 h	Densities; porosity; nutritional property

n/a, not available.

5.3. Speciality Foods

The use of FD is not limited to fruits and vegetables; it has been used to produce dried specialty foods from plant sources such as coffee, tea, and spices [4,7,8,11,58–65].

Coffee and tea are the most popular beverages in the world. FD has been used to produce instant tea due to its ability to retain volatile compounds. Kraujalyte et al. [62] found that instant tea produced by FD had high concentration of volatile compounds (318.65 ng/g), which was comparatively two to five times higher than those produced by other drying methods (68.60 to 143.33 ng/g). In the case of coffee, the content of phenolic acids in coffee beans after FD was reported to increase by 41% more than in fresh green coffee beans [59]. A FD process for coffee has been recently designed using mathematical modeling to optimize energy efficiency and preserve the important flavors and nutrients [9]. Dong et al. [60] conducted a study to observe the effect of FD methods on odor compounds and the aromatic profile of roasted coffee beans. Interestingly, they found that quinic acid, which is a major organic acid that is attributed to coffee quality, was only detected in the sample dehydrated by FD method.

Spices such as garlic and ginger have also been reported to be dehydrated using FD. The effect of freeze-drying shelf temperatures on pore formation of garlic was studied by Sablani et al. [64]. They found that garlic dried at a higher shelf temperature resulted in lower open pore porosity. In addition, the apparent porosity of garlic exponentially increased with the drying time. Ratti et al. [63] investigated the effect of FD on allicin formation capacity. It was found that allicin content decreased with an increase in drying temperature and better retention of allicin formation capacity was obtained from the one dried at 20 °C. In another study, Fante & Noreña [8] found that FD garlic powder demonstrated better quality in terms of color and inulin content, and higher glass transition temperature when compared to forced HAD. High glass transition temperature of 44.9–46.2 °C in freeze-dried garlic powder can be related to low water activity of 0.12 to 0.13. Ginger, a common condiment used in a variety of foods and beverages, is another spice widely dehydrated using FD method [4,7]. FD of ginger led to high retention of gingerols, phenolic content, flavonoids, antioxidant activities, and some volatile compounds [4].

5.4. Nontraditional Source

More recently, there is has been an increasing trend of consumption of nontraditional foods or food from an alternative source. One such nontraditional plant source food is maple syrup. Maple syrup is composed of a mixture of sugars (66% sucrose, 0.4% glucose, and 0.5% fructose), minerals and water, and traces of organic acids, proteins, and polyphenols [66]. Bhatta et al. [67] studied the drying of maple syrup to produce a maple sugar powder. FD of sugar-rich foods is challenging due to high hygroscopicity of simple sugars, the increase in solubility with temperature, a low glass transition temperature of sugars (fructose, glucose, and sucrose; Tg = 5, 31, and 62 °C, respectively) [68], and the stickiness problem in the drying equipment [69]. The dilution of maple syrup from 66 to 20 °Brix was needed to produce a dried maple syrup powder. Such freeze-dried maple sugar powder exhibited an instant-like property as it dissolved within 14s, but showed fair to poor flow characteristics due to cohesiveness nature of sugars. Authors have also suggested the use of glass transition temperatures (related to collapse temperature) for the determination of FD temperatures and online temperature recording with thermocouples for the identification of drying periods [67].

5.5. Generalities about Impact of Freeze-Drying on Biocompounds

Several studies investigated/reviewed the important impact of different drying techniques, including freeze-drying, on the active ingredients and phytochemical contents of fruits, vegetables, and herbs and medicinal plants [70,71]. In particular due to freeze-drying, Marques et al. [57] showed significant losses in vitamin C in tropical fruits (3% to 70% depending on the fruit). Araya-Farias et al. [44] reported 20% loss in vitamin C and in total carotenoids, 35% loss in vitamin E, but only 4% loss in total phenolics when freeze-drying seabuckthorn berries at 20 °C shelf temperature and 30 mTorr vacuum pressure. So, although an excellent choice to preserve plant-based foods, freeze-drying cause some decrease in phytochemical content.

Nevertheless, when compared to other drying techniques, usually freeze-drying is a superior technology. Asami et al. [72] reported that freeze drying preserved total phenolics in marionberries, strawberries, and corn better than air drying. Sablani et al. [70] showed that compared to air drying, freeze drying improved retention of anthocyanins, phenolics, and antioxidant activity during processing of regular versus organic blueberries and raspberries, and in some cases it even increased the concentration of phytochemicals. Reyes et al. [73] also indicated that ascorbic acid content in blueberries was significantly reduced by freeze-drying in any operating condition, while the total polyphenol content was apparently augmented when a vacuum was used (compared to atmospheric pressure), an increase attributed to an improvement in the extractability of polyphenols. To conclude, for vitamin C and phenolic content retention, vacuum freeze drying most of the time gives the best results.

In terms of preserving β-carotene, lycopene, vitamin E, unsaturated oils, and other lipid-based oxidizable bio-compounds in fruits and vegetables, freeze-drying and storage of freeze-dried products should be taken with high consideration since autocatalytic oxidative reactions are accelerated at very low water activities achieved during freeze drying. As an example, Gutierrez et al. [45] indicated that oils from freeze-dried pulps of seabuckthorn berries had a much lower peroxide value than those obtained from air-dried berries, showing that low water activities attained during freeze-drying could damage the quality of lipid-based biocompounds.

6. Pretreatments and Process Intensification

Process intensification with innovative technologies (external to the product) or additional pretreatments (internal) prior to or during freeze-drying are key approaches aimed at efficiently overcoming processing challenges to increase mass transfer or improve product quality.

Chemical, mechanical, and thermal pretreatments have been used to reduce the effect of plant skin hydrophobicity and promote water transport during drying of whole berries. Chemical pretreatment involves immersion of the product in alkaline or acid solutions of oleate esters prior to drying. Alkaline dipping facilitates drying by forming cracks on the fruit surface [74]. However, the high temperature (100 °C) of the chemical solution and the long periods of soaking causes texture degradation and a low level of taste acceptability due to the incorporation of chemical residues in the fruit flesh [75]. Mechanical pretreatments might replace or complement chemical pretreatments, mainly because of the higher acceptability levels [76]. It consists of peeling, abrasion of the surface, puncturing the skin, or cutting the fruit in various shapes [77]. Araya-Farias et al. [44] halved sea-buckthorn berries before freeze-drying in order to produce high quality powders out of this oily, impermeable skinned fruit. Some other pretreatments include exposure to sulphur dioxide and thermal pretreatments such as blanching (immersion in hot water) or steaming [76]. However, blanching may cause the loss of soluble substances like proteins and mineral elements while high temperatures may induce the loss of heat labile substances such as nutrients and vitamins [78].

On the other hand, there has been little research done on freezing pretreatments prior to dehydration methods. Slow freezing helps the formation of large extracellular ice crystals damaging vegetable tissues while rapid freezing promotes intensive nucleation and formation of intracellular small ice crystals [79] and freeze-fractures and cracking in food tissues [80–82]. Water permeability of plant tissues depends on their composition [83], microstructure [84], crystalline or amorphous state of the matrix, and the lipid and glass transitions occurred during cooling or heating the tissues [85,86]. All these reported effects of freezing on vegetable and fruit tissues can certainly be used to induce positive changes in the food microstructure so as to increase drying rates or to improve dried product quality. Individual quick freezing (IQF), a rapid individual freezing of berries in a thin layer at −40 °C for a specified time [87], has been used in cycles with slow thawing in the refrigerator at 4 °C. This mild heat shock (−40 °C to +4 °C), together with the repetition in cycles, led to slight changes in the permeability of the waxy cuticle, sufficient to increase the drying rate [88].

Liquid nitrogen cyclic immersions of blueberries, seabuckthorn berries, and grapes markedly increased the drying kinetics during hot-air, vacuum, and freeze-drying [89]. The initial fruit epidermis thickness decreased between 20% to 50% (depending on the fruit) after 3–5 immersions in liquid N_2. Also, dewaxing of the plant surface was observed after immersions in liquid nitrogen for lowbush (200.33 ± 3.05 to 152.70 ± 0.7 $\mu g/cm^2$) and highbush (227.5 ± 2.12 to 112.17 ± 1.66 $\mu g/cm^2$) blueberry cuticles [32], which explains the significant impact of this pretreatment on mass transfer acceleration during drying.

Dilution of a concentrated product is an easy solution to overcome problems indicated previously when dealing with liquid foods in high sugar/lipid concentrations. However, adding water and afterwards having to taking it out by an expensive method such as freeze-drying is not always an affordable solution unless necessary. Bhatta et al. [67] diluted maple syrup to 20% prior to freeze-drying as a first step of producing ultimate quality maple syrup powders. A pretreatment option for solving quality problems when freeze-drying high concentrated liquids, or solid plant foods having a moisture barrier (i.e., berries), is to provide the freeze-drier with frozen particulate systems instead of a tray of frozen liquid in a block. To achieve this, one simple way is to grind the frozen liquid at ultralow low temperatures. Then, the smaller size frozen particles are freeze-dried. This method was first reported for obtaining freeze-dried powders from plant tissues for botanical analytical use, by grinding the samples under liquid nitrogen, followed by freeze-drying [90]. One of the drawbacks, though, is a wide particle size distribution of the freeze-dried powder.

The more recent freeze granulation technology involves spraying droplets of a liquid slurry or suspension into liquid nitrogen followed by freeze-drying of the frozen droplets [91]. The above-mentioned process is illustrated in Figure 5 (adapted from [92]). The significance of this technology is that the structure and homogeneity of the particles in the slurry or suspension are retained in the granules.

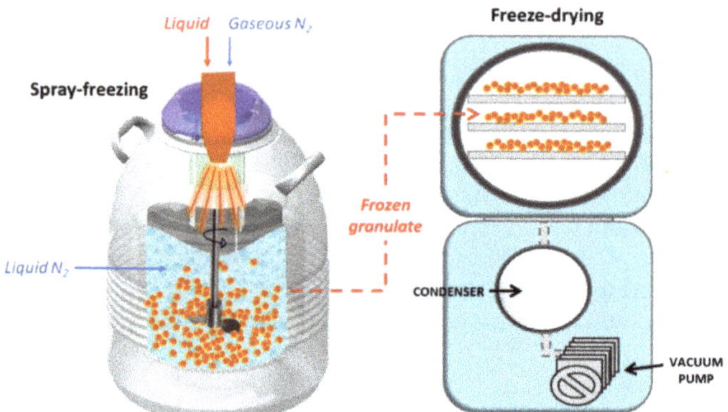

Figure 5. Freeze-granulation process.

Another way of pretreating hard-to-freeze-dry solutions/suspensions/emulsions is by foaming, in the so-called 'foam-mat freeze-drying' process. In general, drying of foamed materials is faster than that of nonfoamed ones. Drying experts have repeatedly pointed to the increased interfacial area of foamed materials as the factor responsible for reduced drying time. However, because density of foamed materials is lower than that of nonfoamed ones and extends from 0.3 to 0.6 g/cm^3, the mass load of the foam-mat dryer is also lower. Thus, shorter drying times should not only offset the reduced dryer load but also increase the dryer throughput. Raharitsifa & Ratti [93] revealed that freeze-drying of foamed apple juice was limited by heat transfer, while for nonfoamed one, by mass transfer. In this study, it was shown that the insulation property characteristic of foams was more significant in slowing down the freeze-drying process than the increased surface area available for mass transfer due to foaming. Although freeze-drying rates were increased by foaming, no practical minimal sample thickness could be found in order to increase freeze-dryer throughput as well. In further experiments, Raharitsifa & Ratti [94] reported that, confirming the glass transition temperature results, at 20 °C storage temperature and in presence of air, nonfoamed freeze-dried apple juice powders collapsed with marked change of color, while foamed freeze-dried products were stable for up to 70 days.

Some technical challenges facing freeze-drying include long residence times, batch operation mode, high operating cost, and energy consumption. It has been already pointed out that any new development to the classical vacuum freeze-drying should aim to improve heat transfer in order to help sublimation and reduce drying times, or to reduce/avoid the use of vacuum so as to decrease costs [3]. In recent years, studies have focused on development of process intensification technologies to resolve some of these issues.

Infrared energy impinges into the exposed material surfaces and propagates through the material to increase thermal energy through molecular vibration, which has relatively lower losses compared to other types of heat sources [95]. Application of infrared radiation in freeze-drying of plant-based foods significantly diminished drying time for sweet potato [96] and apple [97], and enhanced quality characteristics of the final product due to uniform surface heating [98], such as in the case of aloe vera [95], strawberry [99], and banana [100].

Lately, the application of microwave energy to intensify freeze-drying regained attention. Microwave heating has been studied since the 1970s in relation to the acceleration of freeze-drying [3].

The attractive aspect of this heating source is that it is an energy input that not only is essentially unaffected by the dry layers of the material undergoing freeze-drying, but also that is absorbed mainly in the frozen region [101]. Since the frozen region has a high thermal conductivity, microwave energy helps sublimation to decrease freeze-drying times up to 60–75% [102,103]. In addition, when compared to conventional freeze-drying, microwave assisted freeze-drying (MFD) may lead to products of similar/better quality [102,104,105]. Although microwave freeze-drying can offer unique advantages, the inherent problem preventing its commercialization is the difficulty in controlling the final product quality and assuring its uniformity, resulting from corona discharge and nonuniform heating, which cause ice melting and overheating [15]. In an interesting review of microwave assisted freeze-drying of foods, Duan et al. [104] pointed out that to assure a successful implementation of this type of technology in industry, the following challenges should be tackled: operation scale-up, accurate temperature monitoring, appropriate simulation of the MW field distribution, and increase in the knowledge on dielectric properties of foods. Thus, most of the numerous articles published lately considered the impact of microwave freeze-drying on different aspects of the final product quality and uniformity, such as the case for banana [106–109], barley grass [110], potato [108,111,112], mushrooms [113–115], apple [116–118], lettuce stem [65,119,120], okra [121], etc. In a recent review, [122] described an overview of the current developments in microwave assisted freeze-drying of fruits and vegetables, where they concluded the need for other novel nonthermal technologies, such as ultrasounds, high pressure processing, or pulsed electric fields, to improve quality of freeze-dried heat-sensitive fruits and vegetables.

Atmospheric freeze-drying (AFD) also has been getting increased attention recently. In this technique, freeze-drying is done at atmospheric pressure under inert dry gases. Although discovered at the end of the 1950s [123], this freeze-drying process started to pick up scientific interest mainly in the mid 1980s. Due to the lack of vacuum use, the cut off was approximately 34% as compared to vacuum freeze-drying [124]. However, drying times increased 1 to 3 times since the use of atmospheric pressure turns the control of the process from heat to mass transfer, which makes the kinetic extremely slow [3]. Other studies showed that in addition, quality of products was not excellent when atmospheric pressure is used instead of vacuum, since the risk of product collapse increased [125]. In order to accelerate drying kinetics, as well as improve quality issues, AFD has been combined with other techniques such as fluid-bed and spray freeze dryers [123]. As pointed out in this AFD review done by Claussen et al. [123], newer investigations of atmospheric freeze-drying in a fluid bed have looked into a process where a heat pump system is included with the drying system.

More recently, power ultrasound proved to be an effective, nontoxic, and environmentally friendly way to speed the AFD process [126]. Repeated compression–expansion cycles generated by the ultrasound helps to create of micropathways in the solid to ease the vapor flow and microstirring at the-solid fluid interface, reducing the external mass transfer resistance [127]. Moreover, the additional exertion of a mild heating effect increased the interest in the ultrasound-assisted atmospheric freeze-drying of thermally sensitive products.

A compilation of the last 15 years of scientific publications in the area of atmospheric freeze-drying of plant-based foods can be found in Table 3, where their main objectives and conclusions are detailed. Most of these articles deal with improvements of the AFD process by the use of heat pump application to enhance economy aspects or final product quality, or by spout and fluidized beds with and without immersion adsorbents, pulverization by spray, set-up temperature programs, and ultrasound applications as drying strategies to accelerate AFD. Theoretical mathematical modeling [128–130] is an interesting strategy to understand the AFD process and simulate it under diverse conditions, to tackle AFD concrete problems on a solid basis. To end, it was surprising to see that little attention has been paid in some published works to verify that actual freeze-drying instead of air-drying of a frozen product was happening through all the process under atmospheric pressure. Compulsory continuous process humidity determinations and control should be included in future atmospheric freeze-drying studies.

Table 3. Publications on atmospheric freeze-drying of plant-based foods in the last 15 years (in chronological order of publication), where the following abbreviations are used: AFD = atmospheric freeze-drying, CDF = computational fluid mechanics, IAM = immersion in adsorbent material, URIF = uniformly retreat ice front.

Plant-Based Food	Objectives	Conclusions
Apple cubes [131]	To design and build a heat pump-assisted, packed bed AFD closed system and investigate the drying kinetics effect on the quality (rehydration kinetics, shrinkage, color, and antioxidant activity) of apple cubes	Mass diffusion controls the AFD process of apple dewatering at air temperatures below 0 °C. Process temperature had a major impact on final quality. The quality evaluation of apple cubes shows that dried products of AFD at −10 °C have similar rehydration kinetics and hygroscopic properties as the product obtained from vacuum freeze-drying.
Apple cubes [128]	To illustrate the construction and validation process of a CFD model at process temperatures below 0 °C.	CFD results based on film sublimation showed the viability of applying a surface sublimation model to AFD process. CFD results for apple cubes showed a predomination of inertial resistance of porous tissue. True values of tortuosity and internal resistance coefficient are critical for proper process simulation.
Apple cubes [132]	To test a robust and easy modeling tool for predicting AFD performance, designing and scaling-up of the AFD process where shrinkage is taken into consideration, and predictions of AFD drying kinetics can be performed at varying process temperature.	The diffusion model uses an effective diffusivity and activation energy to cover the AFD multicomponent diffusion mechanism. Coupled to shrinkage, the model showed good prediction of the drying kinetics of selected food products in the AFD process. Based on the predictions, it can be concluded that the diffusion model is capable of being applied to simulate AFD process for selected materials at constant and ascending process temperature modes.
Carrot slices [129]	To develop a mathematical model by adopting a sublimation–condensation model for the first stage of freeze-drying, solving the set of equations by fixed-finite-differences. Numerical simulations were carried out to analyze the characteristics of AFD in a fluidized bed dryer.	The complex interface movement in food products was well represented by the method of finite differences, using variable time steps that allowed significant reductions in computer time. The effect of particle size reduction, bed temperature increase, and the incorporation of infrared radiation made it possible to reduce primary drying times. The proposed model of AFD with one-directional mass and energy transport compared well to experimental data.
Carrot parallelepipeds [133]	To study the influence of particle size, freezing rate, air temperature, and mode of energy supply on both the final moisture content and particle shrinkage during AFD in a pulsed fluidized bed.	The air temperature was found to be the most important factor that affected the moisture content, followed by particle size, freezing rate, and type of energy supply. The air temperature was the only factor that affected shrinkage in AFD.
Peas [134]	To study the influence of drying temperatures and ultrasonic intensity on the effective acceleration of AFD rates.	Airborne ultrasound has high potential for improving AFD, as well as other processes that are based on heat and mass transfer rates at low temperatures.
Peas, apple and pinneaple cubes [135]	To use the Weibull model to represent AFD kinetics for different drying temperatures, drying times, approach velocities, products, and particle sizes.	The drying curves for several products obtained using this approach confirmed that AFD is controlled by internal diffusivity. The modified Weibull model adequately described the kinetics with high accuracy and enhanced stability.
Peas [136]	To investigate the application of microwave in AFD of green peas in a porous packed bed and its impact on drying kinetics and product quality.	Drying time was approximately halved when applying microwave radiation of 280 Watt into the process. Process temperature played a major role in product final quality with the lowest temperature being the most favorable to retain color.

Table 3. *Cont.*

Plant-Based Food	Objectives	Conclusions
Apple cubes [137]	To investigate the influence of different drying strategies on the AFD of apples.	A step-up temperature program based on glass transition temperature during AFD process can reduce the drying time by almost half on the premise of keeping product quality.
Apple cubes [130]	To evaluate the feasibility of a simple one-dimensional model to describe the ultrasonic assisted AFD process, as well as to validate such a model in different operating conditions.	A simple one-dimensional model was successfully applied to assess the effect of the ultrasonic application on the AFD kinetics under different conditions. US application is the parameter with the greatest influence on the AFD time and, consequently, is the key factor for the further optimization of the process.
Eggplant cubes [138]	To analyze the ultrasound-assisted AFD process and provide an in silico approach to the industrial process optimization. The URIF model was used to establish the kinetic parameters of the process.	Power ultrasound application significantly reduced the drying time. Air temperature and size of the samples also had a significant impact. The drying kinetics were successfully described using the URIF model. Power ultrasound can also increase the productivity of a tunnel dryer up to four or five times at industrial scale. Despite the benefits that can be envisioned by simulation, some limitations lie on the practice.
Wheat bran and vegetable pieces [139]	To characterize the hydrodynamic behavior of nonfood wheat bran, as potential adsorbent for AFD-IAM in a fluidized bed as well as spout-fluid bed and to study the segregation of binary mixtures of nonfood wheat bran and vegetables at different levels of dryness so as to establish the ideal conditions under which AFD-IAM can be performed without excessively reducing the product size.	Nonfood wheat bran is a promising material to be used as adsorbent. However, because it can be considered a "pseudo-cohesive" powder, potential difficulties in handling the binary mixture may occur when using a fluidized bed in the AFD process. Product density plays a fundamental role in mixing since poor contact between adsorbent and food material was found in the first stages of the AFD process (fluidized bed). Passive and active particle transport mechanisms and blocking effects of floor and roof were proposed to explain the observed behavior together with channeling and collapse cycle, allowing an explanation of the segregation phenomenon in fluidized beds and the mixing process in spout-fluid beds.
Baccharis dracunculifolia D.C. (Bd) plant extract [140]	To verify the influence of fluidized bed atmospheric spray-freeze-drying on the quality of Bd extracts as well as the physical and chemical stability of their main active species during and after the drying.	The main prenylated compounds of Bd are amenable to drying at freezing temperatures, d-mannitol showed an excellent cryoprotectant effect, decreasing the loss of all markers. Also, different powders obtained in the fluidized bed atmospheric spray-freeze-dryer showed adequate morphology, moisture, and excellent pharmaco-technical properties with good process yields. Fluid bed atmospheric spray-freeze-drying is an attractive alternative for processing heat-sensitive and high value-added crop products.
Eggplant cubes [127]	To evaluate the effect of air temperature and velocity, US power, and sample size on vitamin C, total phenolic, and antioxidant capacity of eggplant during US-assisted AFD	Power ultrasound is a promising technology for accelerating the AFD process, but attention must be paid to the optimization of the operating conditions in order to limit the thermal effects of acoustic energy and to ensure the preservation of the nutritional properties of the samples.
Mushroom slices [141]	To evaluate the feasibility of using power ultrasound to improve the AFD of mushroom, considering the kinetic effects and its influence on quality attributes.	Ultrasound represents an interesting means of significantly increasing the drying rate without producing important effects on the final quality of mushrooms.

7. Conclusions

Freeze-drying is widely used to dehydrate the plant-based foods including fruits, vegetables, spices, and even some nontraditional foods. Despite the long processing time and being an expensive drying method, it is preferred for the high final quality. Although some losses in vitamins and other valuable biocompounds can be found after freeze-drying, this type of dehydration method is the best to preserve nutritional qualities when compared to other dehydration methods, especially when operated under vacuum. In addition, quality parameters such as rehydration and porosity of freeze-dried foods are favorable for manufacturing variety of foods such as soup, instant drinks, cakes, etc. More recently, the process intensification of FD with innovative technologies or pretreatments allows overcoming some of the FD processing challenges.

Author Contributions: Conceptualization, C.R.; Formal analysis, C.R. and S.B., Investigation, C.R., T.S.J. and S.B.; Resources, C.R. and T.S.J.; Writing—original draft preparation, C.R. and S.B.; Writing—review and editing, C.R., T.S.J.; Supervision, C.R., Project administration, C.R., Funding acquisition, C.R. and T.S.J. All authors have read and agreed to the published version of the manuscript.

Funding: Natural Sciences and Engineering Research Council of Canada: NSERC RGPIN—2017-04774—operating grant

Conflicts of Interest: The authors declare no conflict of interest.

References

1. FAO/WHO. *Fruit and Vegetables for Health: Report of a Joint FAO/WHO Workshop*; FAO/WHO: Kobe, Japan, 2004.
2. Karam, M.C.; Petit, J.; Zimmer, D.; Djantou, E.B.; Scher, J. Effects of drying and grinding in production of fruit and vegetable powders: A review. *J. Food Eng.* **2016**, *188*, 32–49. [CrossRef]
3. Ratti, C. Hot air and freeze-drying of high-value foods: A review. *J. Food Eng.* **2001**, *49*, 311–319. [CrossRef]
4. An, K.; Zhao, D.; Wang, Z.; Wu, J.; Xu, Y.; Xiao, G. Comparison of different drying methods on Chinese ginger (Zingiber officinale Roscoe): Changes in volatiles, chemical profile, antioxidant properties, and microstructure. *Food Chem.* **2016**, *197*, 1292–1300. [CrossRef]
5. Andriot, I.; Le Quéré, J.-L.; Guichard, E. Interactions between coffee melanoidins and flavour compounds: Impact of freeze-drying (method and time) and roasting degree of coffee on melanoidins retention capacity. *Food Chem.* **2004**, *85*, 289–294. [CrossRef]
6. Ciurzyńska, A.; Lenart, A.; Gręda, K.J. Effect of pre-treatment conditions on content and activity of water and colour of freeze-dried pumpkin. *LWT* **2014**, *59*, 1075–1081. [CrossRef]
7. Ding, S.; An, K.; Zhao, C.; Li, Y.; Guo, Y.; Wang, Z. Effect of drying methods on volatiles of Chinese ginger (Zingiber officinale Roscoe). *Food Bioprod. Process.* **2012**, *90*, 515–524. [CrossRef]
8. Fante, L.; Noreña, C.P.Z. Quality of hot air dried and freeze-dried of garlic (Allium sativum L.). *J. Food Sci. Technol.* **2015**, *52*, 211–220. [CrossRef]
9. Fissore, D.; Pisano, R.; Barresi, A. Applying quality-by-design to develop a coffee freeze-drying process. *J. Food Eng.* **2014**, *123*, 179–187. [CrossRef]
10. Franceschinis, L.; Salvatori, D.M.; Sosa, N.; Schebor, C. Physical and Functional Properties of Blackberry Freeze- and Spray-Dried Powders. *Dry. Technol.* **2014**, *32*, 197–207. [CrossRef]
11. Gümüşay Özlem, A.; Borazan, A.A.; Ercal, N.; Demirkol, O. Drying effects on the antioxidant properties of tomatoes and ginger. *Food Chem.* **2015**, *173*, 156–162. [CrossRef] [PubMed]
12. Hawlader, M.N.A.; Perera, C.O.; Tian, M.; Yeo, K.L. Drying of Guava and Papaya: Impact of Different Drying Methods. *Dry. Technol.* **2006**, *24*, 77–87. [CrossRef]
13. Nindo, C.; Sun, T.; Wang, S.; Tang, J.; Powers, J. Evaluation of drying technologies for retention of physical quality and antioxidants in asparagus (Asparagus officinalis, L.). *LWT* **2003**, *36*, 507–516. [CrossRef]
14. Shishehgarha, F.; Makhlouf, J.; Ratti, C. Freeze-Drying Characteristics of Strawberries. *Dry. Technol.* **2002**, *20*, 131–145. [CrossRef]
15. Wang, Y.; Liu, Y.; Huo, J.; Zhao, T.; Ren, J.; Wei, X. Effect of different drying methods on chemical composition and bioactivity of tea polysaccharides. *Int. J. Boil. Macromol.* **2013**, *62*, 714–719. [CrossRef]
16. Sablani, S.S. Drying of Fruits and Vegetables: Retention of Nutritional/Functional Quality. *Dry. Technol.* **2006**, *24*, 123–135. [CrossRef]

17. Mayer, A.M.; Harel, E. Polyphenol oxidases in plants. *Phytochemistry* **1979**, *18*, 193–215. [CrossRef]
18. Bonnie, T.P.; Choo, Y.M. Oxidation and thermal degradation of carotenoids. *J. Oil Palm* **1999**, *2*, 62–78.
19. Uddin, M.; Hawlader, M.; Ding, L.; Mujumdar, A. Degradation of ascorbic acid in dried guava during storage. *J. Food Eng.* **2002**, *51*, 21–26. [CrossRef]
20. Moraga, G.; Martínez-Navarrete, N.; Chiralt, A. Water sorption isotherms and phase transitions in kiwifruit. *J. Food Eng.* **2006**, *72*, 147–156. [CrossRef]
21. Santos, P.H.S.; Silva, M.A. Retention of Vitamin C in Drying Processes of Fruits and Vegetables—A Review. *Dry. Technol.* **2008**, *26*, 1421–1437. [CrossRef]
22. Ratti, C. *Freeze drying for food powder production, In Handbook of Food Powders: Processes and Properties*; Woodhead Publishing: Cambridge, UK, 2013; pp. 57–84.
23. Patel, S.M.; Doen, T.; Pikal, M.J. Determination of End Point of Primary Drying in Freeze-Drying Process Control. *AAPS PharmSciTech* **2010**, *11*, 73–84. [CrossRef] [PubMed]
24. Tang, X.; Pikal, M.J. Design of freeze-drying processes for pharmaceuticals: Practical advice. *Pharm. Res.* **2004**, *21*, 191–200. [CrossRef] [PubMed]
25. Roos, Y. *Phase Transition in Foods*; Academic Press Inc.: San Diego, CA, USA, 1995; p. 366.
26. Tuso, P.J.; Ismail, M.H.; Ha, B.P.; Bartolotto, C. Nutritional Update for Physicians: Plant-Based Diets. *Perm. J.* **2013**, *17*, 61–66. [CrossRef] [PubMed]
27. Gibson, L.J. The hierarchical structure and mechanics of plant materials. *J. R. Soc. Interface* **2012**, *9*, 2749–2766. [CrossRef] [PubMed]
28. Nguyen, T.K.; Mondor, M.; Ratti, C. Shrinkage of cellular food during air drying. *J. Food Eng.* **2018**, *230*, 8–17. [CrossRef]
29. Ashby, M.F.; Medalist, R.F.M. The mechanical properties of cellular solids. *Met. Mater. Trans. A* **1983**, *14*, 1755–1769. [CrossRef]
30. Waldron, K.; Parker, M.; Smith, A. Plant Cell Walls and Food Quality. *Compr. Rev. Food Sci. Food Saf.* **2003**, *2*, 128–146. [CrossRef]
31. Leary, M. Chapter in Titanium in Medical and Dental Applications. In *Design of Titanium Implants for Additive Manufacturing*; Series in Biomaterials; Woodhead Publishing: Cambridge, UK, 2018; pp. 203–224.
32. Ketata, M.; Desjardins, Y.; Ratti, C. Effect of liquid nitrogen pre-treatments on the osmotic dehydration of blueberries. *J. Food Eng.* **2013**, *116*, 202–212. [CrossRef]
33. Allan-Wojtas, P.; Forney, C.; Carbyn, S.; Nicholas, K. Microstructural Indicators of Quality-related Characteristics of Blueberries—An Integrated Approach. *LWT* **2001**, *34*, 23–32. [CrossRef]
34. Tulloch, A.P. Chemistry of Waxes of Higher Plants. In *Chemistry and Biochemistry of Natural Waxes*; Kolattukudy, P.E., Ed.; Elsevier: Amsterdam, The Netherlands, 1976; pp. 235–287.
35. Riederer, M.; Schönherr, J. Accumulation and transport of (2,4-dichlorophenoxy)acetic acid in plant cuticles. *Ecotoxicol. Environ. Saf.* **1985**, *9*, 196–208. [CrossRef]
36. Baker, E.A. Chemistry and morphology of plant epicuticular waxes. In *The Plant Cuticle*; Cutler, D.F., Alvin, K.L., Price, C.E., Eds.; Academic Press: London, UK, 1982; pp. 139–166.
37. Raharitsifa, N. Freeze-Drying of Fruit Juice: An Optimization Problem. Master's Thesis, Université Laval, Québec, QC, Canada, 2003.
38. Capaccioli, S.; Ngai, K.L. Resolving the controversy on the glass transition temperature of water? *J. Chem. Phys.* **2011**, *135*, 104504. [CrossRef] [PubMed]
39. Khalloufi, S.; Ratti, C. Quality Deterioration of Freeze-dried Foods as Explained by their Glass Transition Temperature and Internal Structure. *J. Food Sci.* **2003**, *68*, 892–903. [CrossRef]
40. Slade, L.; Levine, H. Non-equilibrium behavior of small carbohydrate-water systems. *Pure Appl. Chem.* **1988**, *60*, 1841–1864. [CrossRef]
41. Tsourouflis, S.; Flink, J.M.; Karel, M. Loss of structure in freeze-dried carbohydrates solutions: Effect of temperature, moisture content and composition. *J. Sci. Food Agric.* **1976**, *27*, 509–519. [CrossRef]
42. Karathanos, V.T.; Anglea, S.A.; Karel, M. Structural collapse of plant materials during freeze-drying. *J. Therm. Anal. Calorim.* **1996**, *47*, 1451–1461. [CrossRef]
43. Franks, F. Freeze-drying of bioproducts: Putting principles into practice. *Eur. J. Pharm. Biopharm.* **1998**, *45*, 221–229. [CrossRef]
44. Araya-Farias, M.; Makhlouf, J.; Ratti, C. Drying of Seabuckthorn (Hippophae rhamnoides L.) Berry: Impact of Dehydration Methods on Kinetics and Quality. *Dry. Technol.* **2011**, *29*, 351–359. [CrossRef]

45. Gutierrez, L.-F.; Ratti, C.; Belkacemi, K. Effects of drying method on the extraction yields and quality of oils from quebec sea buckthorn (Hippophaë rhamnoides L.) seeds and pulp. *Food Chem.* **2008**, *106*, 896–904. [CrossRef]

46. Meda, L.; Ratti, C. Rehydration of Freeze-Dried Strawberries at Varying Temperatures. *J. Food Process. Eng.* **2005**, *28*, 233–246. [CrossRef]

47. Seerangurayar, T.; Manickavasagan, A.; Al-Ismaili, A.M.; Al-Mulla, Y.A. Effect of carrier agents on flowability and microstructural properties of foam-mat freeze dried date powder. *J. Food Eng.* **2017**, *215*, 33–43. [CrossRef]

48. Wang, H.; Zhang, S.; Chen, G. Glass transition and state diagram for fresh and freeze-dried Chinese gooseberry. *J. Food Eng.* **2008**, *84*, 307–312. [CrossRef]

49. Guiné, R.P.; Barroca, M.J. Effect of drying treatments on texture and color of vegetables (pumpkin and green pepper). *Food Bioprod. Process.* **2012**, *90*, 58–63. [CrossRef]

50. Que, F.; Mao, L.; Fang, X.; Wu, T. Comparison of hot air-drying and freeze-drying on the physicochemical properties and antioxidant activities of pumpkin (Cucurbita moschata Duch.) flours. *Int. J. Food Sci. Technol.* **2008**, *43*, 1195–1201. [CrossRef]

51. Toor, R.K.; Savage, G.P. Effect of semi-drying on the antioxidant components of tomatoes. *Food Chem.* **2006**, *94*, 90–97. [CrossRef]

52. Rajkumar, G.; Shanmugam, S.; Galvâo, M.D.S.; Leite Neta, M.T.S.; Dutra Sandes, R.D.; Mujumdar, A.S.; Narain, N. Comparative evaluation of physical properties and aroma profile of carrot slices subjected to hot air and freeze drying. *Dry. Technol.* **2017**, *35*, 699–708. [CrossRef]

53. King, V.A.-E.; Liu, C.-F.; Liu, Y.-J. Chlorophyll stability in spinach dehydrated by freeze-drying and controlled low-temperature vacuum dehydration. *Food Res. Int.* **2001**, *34*, 167–175. [CrossRef]

54. Schauss, A.G.; Wu, X.; Prior, R.L.; Ou, B.; Huang, D.; Owens, J.; Agarwal, A.; Jensen, G.S.; Hart, A.N.; Shanbrom, E. Antioxidant Capacity and Other Bioactivities of the Freeze-Dried Amazonian Palm Berry, *Euterpe oleraceae* Mart. (Acai). *J. Agric. Food Chem.* **2006**, *54*, 8604–8610. [CrossRef]

55. Regier, M.; Mayer-Miebach, E.; Behsnilian, D.; Neff, E.; Schuchmann, H.P. Influences of Drying and Storage of Lycopene-Rich Carrots on the Carotenoid Content. *Dry. Technol.* **2005**, *23*, 989–998. [CrossRef]

56. Kyriakopoulou, K.; Pappa, A.; Krokida, M.; Detsi, A.; Kefalas, P. Effects of Drying and Extraction Methods on the Quality and Antioxidant Activity of Sea Buckthorn (Hippophae rhamnoides) Berries and Leaves. *Dry. Technol.* **2013**, *31*, 1063–1076. [CrossRef]

57. Marques, L.G.; Silveira, A.M.; Freire, J.T. Freeze-Drying Characteristics of Tropical Fruits. *Dry. Technol.* **2006**, *24*, 457–463. [CrossRef]

58. Chan, E.; Lim, Y.; Wong, S.; Lim, K.; Tan, S.P.; Lianto, F.; Yong, M. Effects of different drying methods on the antioxidant properties of leaves and tea of ginger species. *Food Chem.* **2009**, *113*, 166–172. [CrossRef]

59. Cheng, K.; Dong, W.; Long, Y.; Zhao, J.; Hu, R.; Zhang, Y.; Zhu, K. Evaluation of the impact of different drying methods on the phenolic compounds, antioxidant activity, and in vitro digestion of green coffee beans. *Food Sci. Nutr.* **2019**, *7*, 1084–1095. [CrossRef] [PubMed]

60. Dong, W.; Hu, R.; Long, Y.; Li, H.; Zhang, Y.; Zhu, K.; Chu, Z. Comparative evaluation of the volatile profiles and taste properties of roasted coffee beans as affected by drying method and detected by electronic nose, electronic tongue, and HS-SPME-GC-MS. *Food Chem.* **2019**, *272*, 723–731. [CrossRef] [PubMed]

61. Dong, W.; Hu, R.; Chu, Z.; Zhao, J.; Tan, L. Effect of different drying techniques on bioactive components, fatty acid composition, and volatile profile of robusta coffee beans. *Food Chem.* **2017**, *234*, 121–130. [CrossRef] [PubMed]

62. Kraujalytė, V.; Pelvan, E.; Alasalvar, C. Volatile compounds and sensory characteristics of various instant teas produced from black tea. *Food Chem.* **2016**, *194*, 864–872. [CrossRef]

63. Ratti, C.; Araya-Farias, M.; Méndez-Lagunas, L.; Makhlouf, J. Drying of Garlic (Allium sativum) and Its Effect on Allicin Retention. *Dry. Technol.* **2007**, *25*, 349–356. [CrossRef]

64. Sablani, S.; Rahman, M.; Al-Kuseibi, M.; Al-Habsi, N.; Al-Belushi, R.; Al-Marhubi, I.; Al-Amri, I. Influence of shelf temperature on pore formation in garlic during freeze-drying. *J. Food Eng.* **2007**, *80*, 68–79. [CrossRef]

65. Wang, Y.; Zhang, M.; Mujumdar, A.S.; Mothibe, J.S. Microwave-Assisted Pulse-Spouted Bed Freeze-Drying of Stem Lettuce Slices—Effect on Product Quality. *Food Bioprocess Technol.* **2013**, *6*, 3530–3543. [CrossRef]

66. Bhatta, S.; Ratti, C.; Stevanovic, T. Impact of drying processes on properties of polyphenol-enriched maple sugar powders. *J. Food Process. Eng.* **2019**, *42*, e13239. [CrossRef]

67. Bhatta, S.; Stevanovic, T.; Ratti, C. Freeze-drying of maple syrup: Efficient protocol formulation and evaluation of powder physicochemical properties. *Dry. Technol.* **2019**, 1–13. [CrossRef]

68. Roos, Y.H. Melting and glass transitions weight carbohydrates of low molecular. *Carbohydr. Res.* **1993**, *238*, 39–48. [CrossRef]

69. Bhandari, B.R.; Datta, N.; Howes, T. Problems Associated With Spray Drying Of Sugar-Rich Foods. *Dry. Technol.* **1997**, *15*, 671–684. [CrossRef]

70. Sablani, S.S.; Andrews, P.K.; Davies, N.M.; Walters, T.; Saez, H.; Bastarrachea, L. Effects of Air and Freeze Drying on Phytochemical Content of Conventional and Organic Berries. *Dry. Technol.* **2011**, *29*, 205–216. [CrossRef]

71. Harguindeguy, M.; Fissore, D. On the effects of freeze-drying processes on the nutritional properties of foodstuff: A review. *Dry. Technol.* **2019**, 1–23. [CrossRef]

72. Asami, D.K.; Hong, Y.-J.; Barrett, D.M.; Mitchell, A.E. Comparison of the Total Phenolic and Ascorbic Acid Content of Freeze-Dried and Air-Dried Marionberry, Strawberry, and Corn Grown Using Conventional, Organic, and Sustainable Agricultural Practices. *J. Agric. Food Chem.* **2003**, *51*, 1237–1241. [CrossRef] [PubMed]

73. Reyes, A.; Evseev, A.; Mahn, A.; Bubnovich, V.; Bustos, R.; Scheuermann, E. Effect of operating conditions in freeze-drying on the nutritional properties of blueberries. *Int. J. Food Sci. Nutr.* **2011**, *62*, 303–306. [CrossRef]

74. Salunkhe, D.K.; Bolin, H.R.; Reddy, N.R. *Storage, Processing and Nutritional Quality of Fruits and Vegetables*, 2nd ed.; CRC Press: Boca Raton, FL, USA, 1991; Volume 1 & 2.

75. Grabowski, S.; Marcotte, M. Pretreatment efficiency in osmotic dehydration of cranberries. In *Transport Phenomena in Food Processing*; Welti-Chanes, J., Velez-Ruiz, F., Barbosa-Cánovas, G.V., Eds.; CRC Press: Boca Raton, FL, USA, 2003; pp. 83–94.

76. Grabowski, S.; Marcotte, M.; Quan, D.; Taherian, A.R.; Zareifard, M.R.; Poirier, M.; Kudra, T. Kinetics and Quality Aspects of Canadian Blueberries and Cranberries Dried by Osmo-Convective Method. *Dry. Technol.* **2007**, *25*, 367–374. [CrossRef]

77. Beaudry, C. Evaluation of Drying Methods on Osmotically Dehydrated Cranberries. Master's Thesis, McGill University, Montreal, QC, Canada, 2001.

78. Branger, A.; Richer, M.; Roustel, S. *Alimentation ET Processus Technologiques*; Educagri: Dijon, France, 2007.

79. Fernández, P.; Otero, L.; Guignon, B.; Sanz, P. High-pressure shift freezing versus high-pressure assisted freezing: Effects on the microstructure of a food model. *Food Hydrocoll.* **2006**, *20*, 510–522. [CrossRef]

80. Chassagne-Berces, S.; Poirier, C.; Devaux, M.-F.; Fonseca, F.; Lahaye, M.; Pigorini, G.; Girault, C.; Marin, M.; Guillon, F. Changes in texture, cellular structure and cell wall composition in apple tissue as a result of freezing. *Food Res. Int.* **2009**, *42*, 788–797. [CrossRef]

81. Pham, Q.T.; Le Bail, A.; Hayert, M.; Tremeac, B. Stresses and cracking in freezing spherical foods: A numerical model. *J. Food Eng.* **2005**, *71*, 408–418. [CrossRef]

82. Shi, A.K.D.X. Thermal Fracture in a Biomaterial During Rapid Freezing. *J. Therm. Stress.* **1999**, *22*, 275–292.

83. Arnous, A.; Meyer, A.S. Comparison of methods for compositional characterization of grape (Vitis vinifera L.) and apple (Malus domestica) skins. *Food Bioprod. Process.* **2008**, *86*, 79–86. [CrossRef]

84. Storey, R.; Price, W. Microstructure of the skin of d'Agen plums. *Sci. Hortic.* **1999**, *81*, 279–286. [CrossRef]

85. Koch, K.; Ensikat, H.-J. The hydrophobic coatings of plant surfaces: Epicuticular wax crystals and their morphologies, crystallinity and molecular self-assembly. *Micron* **2008**, *39*, 759–772. [CrossRef] [PubMed]

86. Goswami, T.; Singh, M. Role of feed rate and temperature in attrition grinding of cumin. *J. Food Eng.* **2003**, *59*, 285–290. [CrossRef]

87. Yang, C.S.T.; Atallah, W.A. Effect of Four Drying Methods on the Quality of Intermediate Moisture Lowbush Blueberries. *J. Food Sci.* **1985**, *50*, 1233–1237. [CrossRef]

88. Yang, A.P.; Wills, C.; Yang, T.C. Use of a Combination Process of Osmotic Dehydration and Freeze Drying to Produce a Raisin-Type Lowbush Blueberry Product. *J. Food Sci.* **1987**, *52*, 1651–1653. [CrossRef]

89. Thromas, V.; Araya-Farias, M.; Ratti, C. Acceleration of the drying process for whole berries by permeabilisation of the epidermis with cryogenic pre-treatments. In Proceedings of the IUFOST World Congress in Food Science and Technology, Cape town, South Africa, 22–26 August 2010.

90. Keys, A.J.; Smith, F.S.; Martin, R.V. Preparation of Freeze-dried Powders of Plant Tissues1. *J. Exp. Bot.* **1963**, *14*, 10–13. [CrossRef]

91. Shanmugam, S. Granulation techniques and technologies: Recent progresses. *BioImpacts* **2015**, *5*, 55–63. [CrossRef]

92. La Lumia, F.; Ramond, L.; Pagnoux, C.; Bernard-Granger, G. Fabrication of homogenous pellets by freeze granulation of optimized TiO2-Y2O3 suspensions. *J. Eur. Ceram. Soc.* **2019**, *39*, 2168–2178. [CrossRef]

93. Raharitsifa, N.; Ratti, C. Foam-Mat Freeze-Drying of Apple Juiceï¿½Part 1: Experimental Data and Ann Simulations. *J. Food Process. Eng.* **2010**, *33*, 268–283. [CrossRef]

94. Raharitsifa, N.; Ratti, C. Foam-Mat Freeze-Drying of Apple Juice Part 2: Stability of Dry Products During Storage. *J. Food Process. Eng.* **2010**, *33*, 341–364. [CrossRef]

95. Chakraborty, R.; Bera, M.; Mukhopadhyay, P.; Bhattacharya, P. Prediction of optimal conditions of infrared assisted freeze-drying of aloe vera (Aloe barbadensis) using response surface methodology. *Sep. Purif. Technol.* **2011**, *80*, 375–384. [CrossRef]

96. Lin, Y.-P.; Tsen, J.-H.; King, V.A.-E. Effects of far-infrared radiation on the freeze-drying of sweet potato. *J. Food Eng.* **2005**, *68*, 249–255. [CrossRef]

97. Nowak, D.; Lewicki, P.P. Infrared drying of apple slices. *Inn. Food Sci. Emer. Technol.* **2004**, *5*, 353–360. [CrossRef]

98. Ciurzyńska, A.; Lenart, A. Freeze-Drying—Application in Food Processing and Biotechnology—A Review. *Pol. J. Food Nutr. Sci.* **2011**, *61*, 165–171. [CrossRef]

99. Shih, C.; Pan, Z.; McHugh, T.; Wood, D.; Hirschberg, E. Sequential Infrared Radiation and Freeze-Drying Method for Producing Crispy Strawberries. *Trans. ASABE* **2008**, *51*, 205–216. [CrossRef]

100. Pan, Z.; Shih, C.; McHugh, T.H.; Hirschberg, E. Study of banana dehydration using sequential infrared radiation heating and freeze-drying. *LWT* **2008**, *41*, 1944–1951. [CrossRef]

101. Sunderland, J.E. An Economic Study of Microwave Freeze-drying. *Food Technol.* **1982**, *36*, 50–52, 54–56.

102. Rosenberg, U.; Bogl, W. Microwave Thawing, Drying, and Baking in the Food Industry. *Food Technol.* **1987**, *41*, 85–91.

103. Peltre, R.P.; Arsem, H.B.; Ma, Y.H. Applications of Microwave Heating to Freeze-drying: Perspective. *AICHE Symp. Ser.* **1977**, *73*, 131–133.

104. Duan, X.; Zhang, M.; Mujumdar, A.S.; Wang, R. Trends in Microwave-Assisted Freeze Drying of Foods. *Dry. Technol.* **2010**, *28*, 444–453. [CrossRef]

105. Barrett, A.H.; Cardello, A.V.; Prakash, A.; Mair, L.; Taub, I.A.; Lesher, L.L. Optimization of Dehydrated Egg Quality by Microwave Assisted Freeze-Drying and Hydrocolloid Incorporation. *J. Food Process. Preserv.* **1997**, *21*, 225–244. [CrossRef]

106. Jiang, H.; Zhang, M.; Mujumdar, A.S.; Lim, R.-X. Comparison of drying characteristic and uniformity of banana cubes dried by pulse-spouted microwave vacuum drying, freeze drying and microwave freeze drying. *J. Sci. Food Agric.* **2014**, *94*, 1827–1834. [CrossRef]

107. Jiang, H.; Zhang, M.; Mujumdar, A.S.; Lim, R.X. Analysis of Temperature Distribution and SEM Images of Microwave Freeze Drying Banana Chips. *Food Bioprocess Technol.* **2013**, *6*, 1144–1152. [CrossRef]

108. Jiang, H.; Zhang, M.; Mujumdar, A.S.; Lim, R.-X. Comparison of the effect of microwave freeze drying and microwave vacuum drying upon the process and quality characteristics of potato/banana re-structured chips. *Int. J. Food Sci. Technol.* **2011**, *46*, 570–576. [CrossRef]

109. Jiang, H.; Zhang, M.; Mujumdar, A.S. Microwave Freeze-Drying Characteristics of Banana Crisps. *Dry. Technol.* **2010**, *28*, 1377–1384. [CrossRef]

110. Cao, X.; Zhang, M.; Mujumdar, A.S.; Zhong, Q.; Wang, Z. Effect of microwave freeze drying on quality and energy supply in drying of barley grass. *J. Sci. Food Agric.* **2018**, *98*, 1599–1605. [CrossRef]

111. Wang, R.; Zhang, M.; Mujumdar, A.S. Effect of Osmotic Dehydration on Microwave Freeze-Drying Characteristics and Quality of Potato Chips. *Dry. Technol.* **2010**, *28*, 798–806. [CrossRef]

112. Wang, R.; Zhang, M.; Mujumdar, A.S. Effects of vacuum and microwave freeze drying on microstructure and quality of potato slices. *J. Food Eng.* **2010**, *101*, 131–139. [CrossRef]

113. Liu, W.C.; Duan, X.; Ren, G.Y.; Liu, L.L.; Liu, Y.H. Optimization of microwave freeze drying strategy of mushrooms (Agaricus bisporus) based on porosity change behavior. *Dry. Technol.* **2017**, *35*, 1327–1336. [CrossRef]

114. Duan, X.; Liu, W.C.; Ren, G.Y.; Liu, L.L.; Liu, Y.H. Browning Behavior of Button Mushrooms during Microwave Freeze Drying. *Dry. Technol.* **2016**, *34*, 1373–1379. [CrossRef]

115. Wang, H.-C.; Zhang, M.; Adhikari, B. Drying of shiitake mushroom by combining freeze-drying and mid-infrared radiation. *Food Bioprod. Process.* **2015**, *94*, 507–517. [CrossRef]

116. Schulze, B.; Hubbermann, E.M.; Schwarz, K. Stability of quercetin derivatives in vacuum impregnated apple slices after drying (microwave vacuum drying, air drying, freeze drying) and storage. *LWT* **2014**, *57*, 426–433. [CrossRef]

117. Li, R.; Huang, L.; Zhang, M.; Mujumdar, A.S.; Wang, Y.C. Freeze Drying of Apple Slices with and without Application of Microwaves. *Dry. Technol.* **2014**, *32*, 1769–1776. [CrossRef]

118. Duan, X.; Ren, G.Y.; Zhu, W.X. Microwave Freeze Drying of Apple Slices Based on the Dielectric Properties. *Dry. Technol.* **2012**, *30*, 535–541. [CrossRef]

119. Pei, F.; Shi, Y.; Gao, X.; Wu, F.; Mariga, A.M.; Yang, W.; Zhao, L.; An, X.; Xin, Z.; Yang, F.; et al. Changes in non-volatile taste components of button mushroom (Agaricus bisporus) during different stages of freeze drying and freeze drying combined with microwave vacuum drying. *Food Chem.* **2014**, *165*, 547–554. [CrossRef]

120. Wang, Y.; Zhang, M.; Mujumdar, A.S.; Mothibe, K.J.; Azam, S.R. Effect of blanching on microwave freeze drying of stem lettuce cubes in a circular conduit drying chamber. *J. Food Eng.* **2012**, *113*, 177–185. [CrossRef]

121. Jiang, N.; Liu, C.; Li, D.; Zhang, Z.; Liu, C.; Wang, D.; Niu, L.; Zhang, M. Evaluation of freeze drying combined with microwave vacuum drying for functional okra snacks: Antioxidant properties, sensory quality, and energy consumption. *LWT* **2017**, *82*, 216–226. [CrossRef]

122. Fan, K.; Zhang, M.; Mujumdar, A.S. Recent developments in high efficient freeze-drying of fruits and vegetables assisted by microwave: A review. *Crit. Rev. Food Sci. Nutr.* **2019**, *59*, 1357–1366. [CrossRef]

123. Claussen, I.C.; Ustad, T.S.; Str⊘Mmen, I.; Walde, P.M. Atmospheric Freeze Drying—A Review. *Dry. Technol.* **2007**, *25*, 947–957. [CrossRef]

124. Wolff, E.; Gibert, H. Développements technologiques nouveaux en lyophilisation. *J. Food Eng.* **1988**, *8*, 91–108. [CrossRef]

125. Lombraña, J.I.; Villarán, M.C. The influence of pressure and temperature on freeze-drying in an adsorbent medium and establishment of drying strategies. *Food Res. Int.* **1997**, *30*, 213–222. [CrossRef]

126. García-Pérez, J.V.; Cárcel, J.A.; Riera, E.; Rosselló, C.; Mulet, A. Intensification of Low-Temperature Drying by Using Ultrasound. *Dry. Technol.* **2012**, *30*, 1199–1208. [CrossRef]

127. Colucci, D.; Fissore, D.; Rosselló, C.; Cárcel, J.A. On the effect of ultrasound-assisted atmospheric freeze-drying on the antioxidant properties of eggplant. *Food Res. Int.* **2018**, *106*, 580–588. [CrossRef] [PubMed]

128. Li, S.; Stawczyk, J.; Zbiciński, I. CFD Model of Apple Atmospheric Freeze Drying at Low Temperature. *Dry. Technol.* **2007**, *25*, 1331–1339. [CrossRef]

129. Bubnovich, V.; Quijada, E.; Reyes, A. Computer Simulation of Atmospheric Freeze Drying of Carrot Slices in a Fluidized Bed. *Numer. Heat Transfer Part A Appl.* **2009**, *56*, 170–191. [CrossRef]

130. Santacatalina, J.; Fissore, D.; Cárcel, J.; Mulet, A.; Garcia-Perez, J.V. Model-based investigation into atmospheric freeze drying assisted by power ultrasound. *J. Food Eng.* **2015**, *151*, 7–15. [CrossRef]

131. Stawczyk, J.; Li, S.; Witrowa-Rajchert, D.; Fabisiak, A. Kinetics of atmospheric freeze-drying of apple. *Dry. Porous Mater.* **2007**, *66*, 159–172.

132. Li, S.; Zbicinski, I.; Wang, H.; Stawczyk, J.; Zhang, Z. Diffusion Model for Apple Cubes Atmospheric Freeze-Drying with the Effect of Shrinkage. *Int. J. Food Eng.* **2008**, *4*. [CrossRef]

133. Reyes, A.; Vega, R.V.; Bruna, R.D. Effect of Operating Conditions in Atmospheric Freeze Drying of Carrot Particles in a Pulsed Fluidized Bed. *Dry. Technol.* **2010**, *28*, 1185–1192. [CrossRef]

134. Bantle, M.; Eikevik, T.M. Parametric Study of High-Intensity Ultrasound in the Atmospheric Freeze Drying of Peas. *Dry. Technol.* **2011**, *29*, 1230–1239. [CrossRef]

135. Bantle, M.; Kolsaker, K.; Eikevik, T.M. Modification of the Weibull Distribution for Modeling Atmospheric Freeze-Drying of Food. *Dry. Technol.* **2011**, *29*, 1161–1169. [CrossRef]

136. Eikevik, T.M.; Alves-Filho, O.; Bantle, M. Microwave-Assisted Atmospheric Freeze Drying of Green Peas: A Case Study. *Dry. Technol.* **2012**, *30*, 1592–1599. [CrossRef]

137. Duan, X.; Ding, L.; Ren, G.-Y.; Liu, L.-L.; Kong, Q.-Z. The drying strategy of atmospheric freeze drying apple cubes based on glass transition. *Food Bioprod. Process.* **2013**, *91*, 534–538. [CrossRef]

138. Colucci, D.; Fissore, D.; Mulet, A.; Cárcel, J.A. On the investigation into the kinetics of the ultrasound-assisted atmospheric freeze drying of eggplant. *Dry. Technol.* **2017**, *35*, 1818–1831. [CrossRef]

139. Coletto, M.M.; Marchisio, D.L.; Barresi, A.A. Mixing and segregation of wheat bran and vegetable pieces binary mixtures in fluidized and fluid-spout beds for atmospheric freeze-drying. *Dry. Technol.* **2017**, *35*, 1059–1074. [CrossRef]
140. Teixeira, C.C.C.; Cabral, T.P.D.F.; Tacon, L.A.; Villardi, I.L.; Lanchote, A.D.; De Freitas, L.A.P. Solid state stability of polyphenols from a plant extract after fluid bed atmospheric spray-freeze-drying. *Powder Technol.* **2017**, *319*, 494–504. [CrossRef]
141. Carrión, C.; Mulet, A.; Garcia-Perez, J.V.; Cárcel, J.A. Ultrasonically assisted atmospheric freeze-drying of button mushroom. Drying kinetics and product quality. *Dry. Technol.* **2018**, *36*, 1814–1823. [CrossRef]

Article

The Impact of Freeze-Drying Conditions on the Physico-Chemical Properties and Bioactive Compounds of a Freeze-Dried Orange Puree

Marilú A. Silva-Espinoza [1], Charfedinne Ayed [2], Timothy Foster [2], María del Mar Camacho [1] and Nuria Martínez-Navarrete [1,*

[1] Food Technology Department, Food Investigation and Innovation Group, Universitat Politècnica de València, Camino de Vera s/n, 46022 Valencia, Spain; masiles@doctor.upv.es (M.A.S.-E.); mdmcamvi@tal.upv.es (M.d.M.C.)

[2] Department of Food, Nutrition and Dietetics, School of Biosciences, University of Nottingham, Sutton Bonington Campus, Loughborough LE12 5RD, UK; Charfedinne.Ayed@nottingham.ac.uk (C.A.); Tim.Foster@nottingham.ac.uk (T.F.)

* Correspondence: nmartin@tal.upv.es

Received: 28 November 2019; Accepted: 24 December 2019; Published: 30 December 2019

Abstract: Fruits are essential for a healthy diet, as they contribute to the prevention of cardiovascular diseases and some cancers, which is attributed to their high bioactive compound content contributing to their antioxidant capacity. Nevertheless, fruits have a short shelf life due to their high-water content, and freeze-drying is a well-known technique to preserve their nutritive quality. However, it is an expensive technology, both due to the use of low pressure and long processing time. Therefore, an optimisation of variables such as the freezing rate, working pressure and shelf temperature during freeze-drying may preserve fruit quality while reducing the time and costs. The impact of these variables on colour, porosity, mechanical properties, water content, vitamin C, total phenols, β-carotene, and antioxidant activity of a freeze-dried orange puree was evaluated. The results showed a great impact of pressure and shelf temperature on luminosity, chroma and water content. Vitamin C and β-carotene were more preserved with higher shelf temperatures (shorter times of processing) and lower pressure, respectively. The optimum freeze-drying conditions preserving the nutrients, and with an interesting structural property, perceived as a crunchy product by consumers, are low pressure (5 Pa) and high shelf temperature (50 °C).

Keywords: vitamin C; total phenols; total carotenoids; antioxidant activity; colour; mechanical properties; pressure; shelf temperature; freezing rate

---

1. Introduction

It is known that there is interest in the consumption of fruits, as they are recommended as components of a healthy diet due to their contribution to the prevention of some diseases when they are consumed in adequate quantity [1,2]. This effect is attributed to their high content of bioactive compounds such as phytochemicals, some vitamins and fibre [3]. In particular, the orange and its derived products are a rich source of flavonoids (mainly hesperidin), carotenes, and vitamin C, with concentrations in the range of 15–238.8 mg, 182–198 µg and 43.5–50 mg/100 g edible fruit, respectively [4–9]. In fact, an average orange would contribute 80% of the RDA (recommended daily allowance) of vitamin C [10]. However, fruits have two main problems that affect their continuous availability, which are seasonality and short shelf life. Dehydration is one of the most common techniques used to preserve food. In addition, it also entails a reduction in the volume and weight of the product, which facilitates its transport and handling [11].

Freeze-drying is a dehydration technique based on the sublimation of the water present in a product, which results in a reduction of water activity and therefore the related deterioration processes to which a food is subjected [12]. The product is frozen in order to be subjected to vacuum pressure with the consequent sublimation and desorption of the water. Freeze-drying operates at low temperatures, which contributes to preserve characteristics such as taste, colour or appearance and to minimize the degradation of thermolabile compounds, many of them responsible for the aromas and nutritional value of the fruits. Thus, the final freeze-dried product is high quality as compared with other techniques of dehydration [13].

Despite the improved microbiological stability of the final product, the chemical and physical attributes may be sometimes compromised. On the one hand, the high porosity and the low water content of the freeze-dried products make the interaction between the solutes and the oxygen at the end of the process more accessible. In this way, the oxidation of bioactive compounds, such as vitamin C, phenols or carotenoids may be promoted. On the other hand, the physical problems are related to the glass transition of the amorphous matrix, which is usually developed during the freeze-drying process. Above the glass transition temperature (Tg), the change from the more stable glassy state to the rubbery state occurs [14]. Freeze-dried fruit pulps, as sugar-rich foods, have a low Tg value in the range of 5–15 °C [15,16]. For this reason, they present collapse and other structural problems related to stickiness and caking, which begin to be developed about 20 °C above Tg [14]. A usual way to delay these problems is the incorporation of high molecular weight biopolymers that contribute to an increase in Tg, or that exert a steric role [17,18].

The disadvantage of freeze-drying is its high cost, due to the long process times and the energy cost related to the vacuum stage. For this reason, it has only been widely used to obtain products with high value added, as occurs in the pharmaceutical industries as well as in some specific food industries, such as rehydratable coffee. However, given the high sensory and functional value of fruits, associated with their high content of bioactive compounds, freeze-drying can be a niche opportunity in this case. In this sense, the technique can provide different food formats, among them, a crunchy fruit product with good consumer acceptance as a snack [18]. Despite adequate optimization of the process conditions contributing to reduce the duration of the process, several reports have indicated that both the freezing and the drying variables, such as the freezing rate or the working pressure and shelf temperature during the drying step, may affect the quality of the obtained product [18–24]. As regards the impact of increasing the shelf temperature, a study carried out with grapefruit puree indicated a decrease of more than 50% in drying time when increasing the temperature up to 40 °C, without a great impact on aspects such as colour, texture or vitamin C content [25]. Nor was an effect observed on the vitamin C content when a mandarin juice was freeze-dried at 40 °C compared to that processed at room temperature [18]. Nevertheless, the shelf temperature should not exceed either the collapse temperature or that which could cause damage to the thermolabile compounds of interest.

In this study, the impact of freeze-drying conditions on the quality of a freeze-dried orange puree with added gum Arabic and bamboo fibre was evaluated. Two freezing rates (conventional and blast freezer), three different shelf temperatures (30, 40, 50 °C) and two working pressures (5 and 100 Pa) were combined. The quality indices measured were the water content, colour, porosity, mechanical properties, vitamin C, carotenoids and phenolic content, as well as the total antioxidant capacity.

2. Materials and Methods

2.1. Raw Materials

Oranges (*Citrus x sinensis* cultivar Navel) used in this study were selected by subjective visual inspection based on a similar weight and size colour homogeneity and good physical integrity (absence of external physical damage). They were bought in October 2019 from a local supermarket in the city of Valencia (Spain) and immediately processed. Carriers used to obtain the dehydrated orange samples were gum Arabic (GA, Scharlab, Sentmenat, Spain) and bamboo fibre (BF, VITACEL®, Rosenberg, Germany).

2.2. Freeze-Drying Processing

Oranges were washed, peeled, cut and triturated in a bench top electrical food processor for 40 s at speed 4 (2000 rpm) followed by 40 s at speed 9 (9100 rpm) (Thermomix TM 21, Vorwerk, Spain). The orange puree was mixed for 10 min at speed 3 (1000 rpm) with (5 g GA + 1 g BF)/100 g orange puree as to ensure the physical stability of the dried product [26]. The formulated orange puree (FOP) was distributed in 10.5 × 7.8 cm aluminum plates of 0.5 cm thickness. Samples were immediately frozen at two different freezing rates (FR): slow freezing (FR-S) in a conventional freezer (Liebherr Mediline LGT 2325, Liebherr, Baden-Wurtemberg, Germany) for 48 h and fast freezing (FR-F), where the samples were frozen for 3 h at −38 °C in a blast freezer (Hiber RDM051S, Hiber, Cernusco sul Naviglio, Italy), and then stored at −45 °C in the conventional freezer for at least 24 h. Frozen samples were dried in a freeze-drier (Telstar Lyoalfa-6, Telstar, Terrassa, Spain) at different pressures (P) in the chamber and shelf temperatures (T). Twelve different conditions were studied (Table 1). The shelf temperature conditioned the drying time, this being 25 h at 30 °C, 7 h at 40 °C and 6 h at 50 °C. The time was selected based on preliminary experiments to be enough to achieve a water content lower than 4%. At these conditions, the physical stability of the formulated puree was known to be guaranteed, as no structural collapse was observed.

Table 1. Sample and conditions code according the 12 different freeze-drying conditions studied.

Sample Code	Shelf Temperature (T)			Pressure (P)		Freezing Rate (FR)	
	30 °C	40 °C	50 °C	5 Pa (P_5)	100 Pa (P_{100})	Slow (S)	Fast (F)
S_30_P_5	X			X		X	
F_30_P_5	X			X			X
S_30_P_{100}	X				X	X	
F_30_P_{100}	X				X		X
S_40_P_5		X		X		X	
F_40_P_5		X		X			X
S_40_P_{100}		X			X	X	
F_40_P_{100}		X			X		X
S_50_P_5			X	X		X	
F_50_P_5			X	X			X
S_50_P_{100}			X		X	X	
F_50_P_{100}			X		X		X

2.3. Water Content

The water content (x_w, g water/100 g product) of FOP was determined using the AOAC method [27]. The sample was dried in a vacuum oven (Selecta®, Vaciotem-T, J.P. Selecta S.A., Barcelona, Spain) at 60 ± 1 °C under $P < 100$ mm Hg until constant weight (XS204 DeltaRange®, Mettler Toledo, Switzerland). For the freeze-dried puree, an automatic Karl Fisher titrator (Mettler Toledo, Compact Coulometric Titrator C10S, Worthington, OH, USA) was used to obtain the water content. Triplicates were performed in each case.

2.4. Mechanical Properties

The mechanical behaviour of the freeze-dried puree was registered using a texture analyser (TA-XT2i, Stable Micro Systems, Godalming, UK). Portions of 20 × 20 mm of the freeze-dried puree were compressed using a cylindrical probe of 10 mm diameter, applying a strain of 80% with a test speed of 1 mms^{-1}. Six replicates were performed per sample. The parameters analysed in the test were the force required to fracture the sample (Fracture force, F_f), expressed in Newtons, and the slope

(S, N/mm) of the curve in the linear zone prior to fracture point, related to the sample resistance to deformation (rigidity) [28].

2.5. Colour Measurements

The CIE L*a*b colorimetric space was considered to characterize the colour [29]. A colorimeter (Minolta, CM 3600D, Japan) was used to measure the colour of the surface of the freeze-dried puree, taking the system observer 10° and illuminant D65 as reference. Colour coordinates, L*, a*, b*, were obtained for each freeze-dried puree. From them, the hue angle (h*, Equation (1)) and chroma or saturation (C*, Equation (2)) were obtained. When total colour differences (ΔE*) were calculated, Equation (3) was used. Measurements were carried out with the specular component excluded. Six replicates were performed per sample.

$$h^* = \arctan(b^*/a^*) \tag{1}$$

$$C^* = (a^{*2} + b^{*2})^{0.5} \tag{2}$$

$$\Delta E^* = ((\Delta L^*)^2 + (\Delta a^*)^2 + (\Delta b^*)^2)^{1/2} \tag{3}$$

2.6. Porosity

True density (ρ) and apparent density (ρ_a) were obtained in order to obtain the porosity (ε, %). True density was calculated based on the sample composition (Equation (4)). Portions of the cakes were obtained with a punch of 22 mm diameter, and were exactly measured in height and diameter with a calliper. Apparent density of each portion was calculated from the weight (m, g; XS204 DeltaRange®, Mettler Toledo, Switzerland) and corresponding volume (V, cm$_3$) (Equation (5)). The porosity was calculated from Equation (6).

$$\rho = \frac{1}{\frac{x_w}{\rho_w} + \frac{x_{CH}}{\rho_{CH}}} \tag{4}$$

where x_w and x_{CH} are the mass fractions of the two main components of each sample (water and carbohydrates, respectively, x_w was determined as described in Section 2.3, and x_{CH} by difference); ρ_w and ρ_{CH} are their densities (ρ_{CH} = 1.4246 g/cm^3, ρ_w = 0.9976 g/cm^3 [30]).

$$\rho_a = \frac{m}{V} \tag{5}$$

$$\varepsilon(\%) = 100\frac{\rho - \rho_a}{\rho} \tag{6}$$

2.7. Total Polyphenolic Compounds

The extraction of total phenolic compunds (TP) was carried out according to Tomás-Barberán et al. [31] with minor modifications. FOP (2.5 g) or freeze-dried puree (0.5 g) were mixed with 9 mL of methanol:water (70:30) using a magnetic multi-stirrer at 200 rpm (JEIO TECH Lab Companion MS-51M, JEIO TECH Lab Companion, Seoul, Korea) under darkness and at room temperature for 30 min. The homogenates were centrifuged at 11,515× *g* at 4 °C for 10 min (GYROZEN Co., 1236R, GYROZEN, Daejeon, Korea). The supernatant was collected and analysed as to TP using the Folin–Ciocalteu method, which was adapted from Benzie et al. [32] with some modifications as described by Selvendran et al. [33]. The TP content was calculated as mg of gallic acid equivalents (GAE)/100 g dry basis (db) sample, using a standard curve in the range of 0–1000 ppm of gallic acid (Sigma-Aldrich, Saint Louis, MO, USA). In this study, all bioactive compounds were referred to the percentage (%) of the corresponding bioactive compound preserved in the freeze-dried puree (FDP) in reference to the FOP, calculated based on Equation (7). This test was done in triplicate for each sample.

$$PBc\ (\%)\ =\ \frac{Bc_{FDP}}{Bc_{FOP}} \times 100 \tag{7}$$

where PBc (%) is the percentage of the corresponding bioactive compound preserved; Bc_{FDP} is the bioactive compound content in the freeze-dried puree (mg/100 db); and Bc_{FOP} is the bioactive compound content in the formulated orange puree (mg/100 g db).

2.8. Antioxidant Activity

The antioxidant activity (AOA) was determined with the DPPH and FRAP tests. The methanolic extract obtained for the quantification of TP was used to this end. DPPH was carried out according to Brand-Williams et al. [34] with minor modification. For these samples, the steady state of the reaction was reached at 15 min, when the absorbance at 515 nm was measured again. The FRAP test was carried out according to Benzie et al. [32]. The results for both methods were converted to mmol Trolox equivalents/100 g db freeze-dried puree. The AOA was also expressed as the percentage (%) of this activity preserved in the FDP in reference to the FOP (Equation (7)). Three replicates were performed per sample.

2.9. Vitamin C

Total vitamin C content (VC) was determined by the reduction of dehydroascorbic acid to ascorbic acid (AA) using high-performance liquid chromatography (HPLC) (Jasco, Italy). The reduction was carried out by mixing 0.5 g of FOP or 0.075 g of each of the 12 freeze-dried puree samples with 2 mL of a 20 g/L DL-dithiothreitol solution (Scharlab, Spain) for 2 h at room temperature and under darkness [35,36]. The extraction of the mixture was carried out according to Xu et al. [37]. The HPLC conditions were: Kromaphase100-C18, 5 mm (4.6 × 250 mm) column (Scharlab SL); mobile phase 0.1% oxalic acid, volume injected 10 µL, flow rate 1 mL/min, detection at 243 nm (detector UV-visible MD-1510, Jasco, Cremella, Italy) at 25 °C. A standard solution of L (+) ascorbic acid (Scharlab SL, Sentmenat, Spain) in the range of 5–200 ppm was prepared. The VC content was calculated as mg AA/100g db sample and the percentage (%) of this bioactive compound preserved in the FDP in reference to the FOP was calculated (Equation (7)). Three replicates were performed per sample.

2.10. β-Carotene

The extraction of β-carotene (BC) was performed using the method of Olives et al. [38] with some modifications. FOP (0.8 g) or freeze-dried puree (0.2 g) were mixed with 9 mL of hexane/ethanol/acetone (50:25:25, *v/v/v*) using a magnetic multi-stirrer at 200 rpm (JEIO TECH Lab Companion MS-51M, Korea), under darkness and at room temperature for 30 min. The homogenates were centrifuged at 11,515× *g* at 4 °C for 10 min (GYROZEN Co., 1236R, Daejeon, Korea). Distilled water was added to the supernatant (10 mL distilled water/100 mL supernatant) and was manually stirred for 2 min. The absorbance of the upper layer was measured at 446 nm (spectrophotometer V-1200 VWR, VWR, Radnor, PA, USA) [38]. The BC was calculated as mg BC/100 g db sample using a β-carotene (Dr. Ehrenstorfer, Augsburg, Germany) calibration curve in the range of 0.5–7 ppm. The BC was referred to the percentage (%) of this bioactive compound preserved in the FDP in reference to the FOP (Equation (7)). Three replicates were performed per sample.

2.11. Statistical Analysis

Data were subjected to Partial Least Squares Regression (PLS-R) and a three way analysis of variance (ANOVA) using Tukey's HSD test to establish the significant effect of shelf temperature, pressure and freezing rate on the parameters studied, with 95% confidence interval, by using XLSTAT statistical and data analysis solution (Addinsoft, 2019, Long Island, NY, USA). F-Values obtained with the ANOVA were also considered to identify the most important factors. Furthermore, a Pearson's correlation analysis between antioxidant capacity and the bioactive compounds was carried out.

3. Results and Discussion

All the results obtained are detailed in the Supplementary Materials (Tables S1–S3). The most relevant aspects are detailed below.

3.1. Physicochemical Characterization

The experimental results of the colour characterization, mechanical properties, porosity, and water content of the freeze-dried purees obtained under each of the 12 studied conditions were processed by PLS-R (Figure 1 and Table S4). Axis 1 (t1) mainly represents the impact of pressure on the qualitative explanatory variables (Y), while axis 2 (t2) represents the impact of shelf temperature on Y. The vectors of freezing rate are in the inner circle, which indicates that in general, these factors are not significantly correlated with the different studied properties of the samples. The variables Y that are significantly affected by a specific dependent variable (X) are circled in red, blue, orange and green.

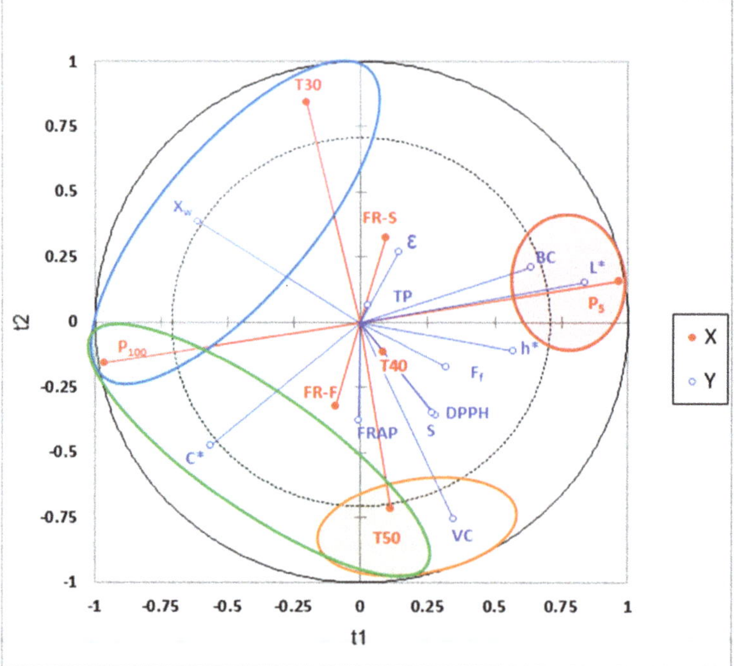

Figure 1. PLS-R projections of freeze-dried orange puree at a range or pressure, temperature and rate of freezing conditions. The dependent variables (X) are projected in red and the qualitative explanatory variables (Y) are projected in blue. The variables Y significantly correlated with a specific dependent variable are circled in red, blue, orange and green. Variables X: T30, T40, T50: shelf temperature applied during freeze-drying at 30, 40 and 50 °C, respectively; P_5 and P_{100}: working pressures of 5 Pa and 100 Pa respectively; FR-S: slow freezing rate and FR-F: fast freezing rate. Variables Y: percentage (Equation (7)) of TP: total phenols, VC: vitamin C, BC: beta carotene, antioxidant activity (FRAP and DPPH); F_f: fracture force (N); S: slope related to rigidity (N/mm); Xw: water content (g water/100 g sample); ε: porosity (%).

3.1.1. Colour

The PLS–R revealed a good correlation between pressure and L*, C*, h* values. They can be observed according to axis 1: P_5 projected on the positive side while P_{100} is projected on the negative side. The external and near position of P_5 to L* and P_{100} to C* denotes a significant effect of pressure on colour attributes, so that low pressure leads to a high L* value (see red circle in Figure 1) while high pressure leads to a high C* value. Despite h* also being positively projected on axis 1 and so affected by pressure, its projection stays in the inner circle of the PLS-R although near the external limit. This indicates a lowered impact of the pressure on the hue angle when compared with that observed for C* and L*. It can also be notice that the projection of C*on the PLS-R, in the negative part of axis 1 and axis 2, is higher when the interaction between P_{100} and T50 is considered (see green circle in Figure 1). The shorter freeze-drying process carried out at 50 °C contributes to promoting freeze-dried products with a higher value of chroma. In addition, C* is projected in the same side of the fast freezing rate (FR-F), the last one being in the inner circle (poorly correlated according to axis 2).

These observations were confirmed by the ANOVA, as values of L*, C* and h* of the samples were significantly affected by working pressure ($p < 0.05$). Further, C* was also affected by the interaction between shelf temperature and freezing rate ($p < 0.05$), but with a low F-Value (7.86). Taking into account the significances shown by the PLS-R analysis and the F-Values of the ANOVA, Figure 2 was constructed, showing L* and C* values of the samples obtained at the different pressure and shelf temperature and considering the mean value at both freezing rates. When working with the highest pressure during freeze-drying (P_{100}), the samples showed lower values of L* and higher C*, which means a darker and saturated colour. In this case, the chroma is specially enhanced at a higher temperature in the freeze-drier shelves, either 40 °C or 50 °C (Figure 2, $p < 0.05$). The hue angle, with values between 80.3 and 82.6, showed a lower value when working with high pressure and freeze-drier shelves temperature below 50 °C ($p < 0.05$). Similar results have been reported by Hammami et al. [22], who noted a slight L* decrease when working at higher pressures ($P > 108$ Pa) for strawberries pieces, which was related to the pronounced shrinkage observed under these conditions. Different authors found that the operating working pressure should be lower than 50 Pa to avoid shrinkage for strawberries pieces and banana slides [22]. With regards to shelf temperature, increasing temperature may cause some slight sugar browning reactions like non-enzymatic or Maillard reactions, which means a reinforcement of the colour shown by the increase in C*. Nevertheless, it is better detected in shrunken samples due to its different optical light reflection capacity [22].

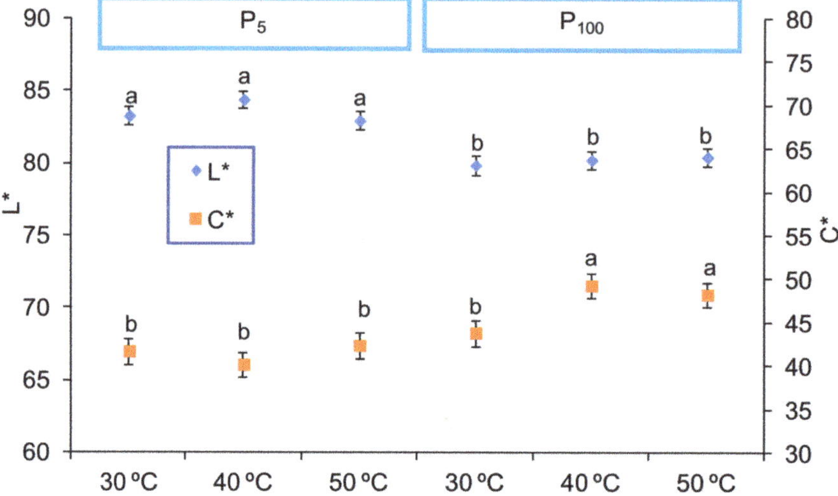

Figure 2. Values (mean and Tukey's HSD) of L* in left axis and C* in right axis of the freeze-dried purees according to the interaction between shelf temperature (30, 40 or 50 °C) and pressure (P₅: 5 Pa and P₁₀₀: 100 Pa) factors. Different letters for each attribute indicate different homogeneous groups for the Temperature*Pressure interaction ($p < 0.05$). Data of both freezing rates are considered in the mean values.

According to the PLS-R and the ANOVA, pressure and shelf temperature were the factors that had a significant impact on the colour. The total differences in colour were calculated in order to evaluate the impact of both factors on the colour. Regarding the shelf temperature, values of ΔE^* between 1.11 and 5.51 were obtained, while the range was 4.45–9.98 for the pressure influence. According to Bodart et al. [39], total differences in colour are not obvious to the human eyes when $\Delta E^* < 1$, minor colour differences could be appreciated by the human eye depending on the hue when $1 < \Delta E^* < 3$ and visually obvious changes for human eyes occur when $\Delta E^* > 3$. Therefore, it seems that pressure has a greater effect than the other factors on the colour.

3.1.2. Structure

Freeze-dried purees showed high porosity, typical of freeze-dried products, with values ranging from 86.4–87.4%. The PLS–R indicated a correlation between FR-S and porosity (mainly projected in the positive side of axis 2, Figure 1), which means that a slow rate of freezing leads to bigger ice crystals and so a bigger expansion of gas cells in the structure (higher porosity) during the freeze-drying process. Nevertheless, they are in the inner circle, which means the effect is not significant. In fact, according to ANOVA, porosity of the samples was not significantly affected by any factor evaluated ($p > 0.05$).

As a result of the mechanical analysis, the force versus distance curves were obtained for each sample freeze-dried under each of the 12 conditions (Figure 3). The linear regression of the first part of the curve was taken to calculate the slope before F_f, related to the rigidity of the sample. As can be observed in the PLS-R (Figure 1), factors T50 and rigidity are both partially correlated and projected on the negative side of axis 2, which means that a higher temperature leads to an increase in the rigidity of the freeze-dried samples. However, because the rigidity is projected in the inner circle, it indicates a moderate positive correlation. In regards to the ANOVA results, the shelf temperature had a significant effect on the rigidity (slope) of the freeze-dried samples ($p < 0.05$); however, its low F-Value (8.65) makes it possible to confirm its low significance.

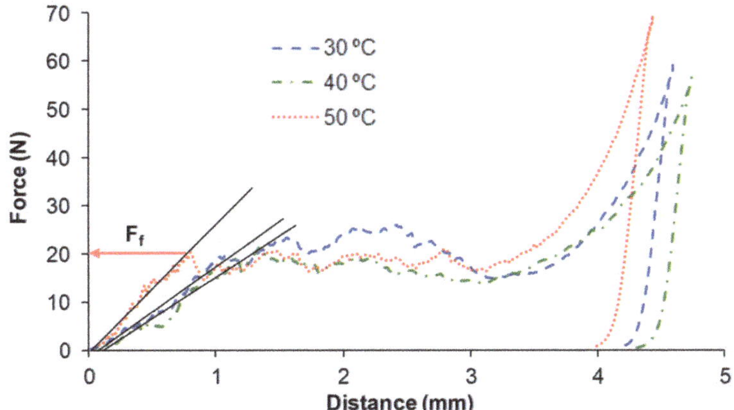

Figure 3. Examples of force–distance curves obtained from the freeze-dried purees frozen at a slow rate and freeze-dried at 5 Pa, and with different shelf temperatures (30, 40 and 50 °C). F_f: fracture force.

Values of the slope 18–26 N/mm were obtained when heating the freeze-drier shelves to 50 °C, as compared to 11–18 N/mm values obtained at 30 and 40 °C. As a greater slope indicates less deformation of the sample by exerting an effort on it, it can be concluded that freeze-drying heating the shelves to 50 °C promotes the mechanical rigidity of the freeze-dried samples before they fracture. This can be considered desirable as it would be related to a higher mechanical resistance of the sample during its handling before consumption.

Fracture force of all the samples varied between 12.1 ± 1.6 N and 19 ± 5 N. The PLS-R revealed a moderate positive correlation between fracture force and both lower pressure (P_5) (both projected in the positive side of axis 1) and T50 (both projected in the negative side of axis 2). This means that working with lower pressure and higher temperature seems to promote a freeze-dried puree that is more resistant to fracture. According to the ANOVA, neither the temperature nor the freezing rate showed a significant effect on the fracture force ($p > 0.05$). The pressure did show $p < 0.05$, but again, with a very low *F*-Value (7.17).

The water content of the samples ranged between 2.2 ± 0.3 and 4.2 ± 0.2 g water/100 g sample. The PLS-R indicated a high correlation between water content (%), high pressure (P_{100}) and low temperature (T30), circled in blue colour in Figure 1. In fact, this statement was confirmed by the ANOVA that indicated a significant effect on the final water content of the freeze-dried samples by the shelf temperature and the pressure ($p < 0.05$). In this way, drying at the highest pressure (100 Pa) and the lowest shelf temperature (30 °C) promoted samples with the highest water content. On the other hand, Figure 1 shows a clear negative correlation between water content and fracture force, which means the smaller the water content of the freeze-dried samples, the greater the mechanical resistance to fracture. In this case, the final water content of the samples was lower than 4.2%. However, it seems that small changes outside this range, related to working conditions, will have an important impact on the mechanical properties of the sample.

3.2. Bioactive Compounds

The quantification of total phenols, vitamin C, β-carotene, and antioxidant activity evaluated by two methods (FRAP and DPPH), was carried out in the FOP and after being freeze-dried for each of the 12 conditions evaluated.

The PLS-R indicated that TP is not projected on the bi-plot (Figure 1), which means that it is relatively less impacted than the other bioactive compounds (TP vector very close to the origin 0.0). In fact, according to the ANOVA, the TP was not affected by working pressure and freezing rate ($p > 0.05$) and was better preserved when the freeze-drying was carried out at 30 or 50 °C as compared to 40 °C

($p < 0.05$), despite the temperature factor also having a low F-Value (8.07). In fact, the most marked difference was observed with the factor temperature, which was 5% of preservation between 30 and 40 °C (Figure 4). No significant interactions between the factors were observed ($p > 0.05$).

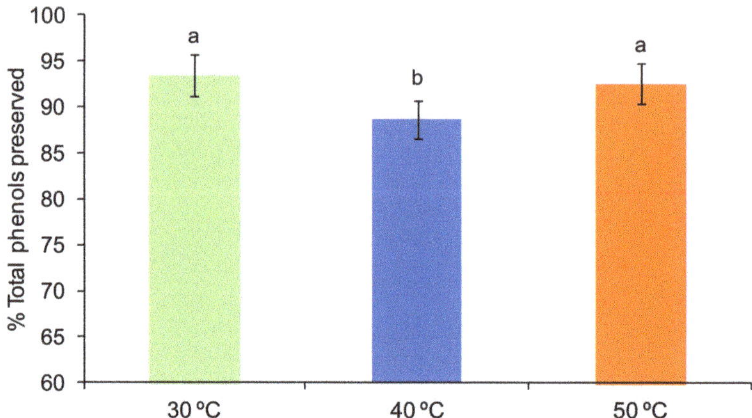

Figure 4. Percentage (%) of preserved total phenols (mean and Tukey's HSD) of samples according to the shelf temperature factor (30, 40 or 50 °C). Different letters indicate different homogeneous groups for the shelf temperature factor ($p < 0.05$). Data of both freezing rates and both pressures are considered in the mean values.

Figure 4 shows TP preservation by shelf temperature, considering both freezing rates and pressures in the mean values. It seems that TP may be affected by the ratio of time and temperature of processing, as reported by other authors [40]. In this case, mild 40 °C heating for more than 7 h seems to compromise TP preservation.

The presence of VC in the final product is used as a reference of high nutritional quality for the different industrial processes, due to its relative instability to heat, oxygen and light [41,42]. The impact of temperature on VC can be clearly observed on the PLS-R according to axis 2 (Figure 1). Vector T50 and T40 are projected in the same direction (negative side of axis 2), while T30 is anti-correlated to them (projected on the positive side of axis 2). This confirms that a higher shelf temperature along the process preserved the vitamin C of the samples. According to the ANOVA, it can be confirmed that VC was affected by the shelf temperature and the pressure during freeze-drying (Figure 5, $p < 0.05$). A significant interaction between both factors indicated that heating the freeze-drier shelves to 40 or 50 °C promoted samples with higher vitamin C content than those freeze-dried at 30 °C. Despite VC being reported to have thermal stability [41], the length of time required when the freeze-drying is carried out at 30 °C (25 h) may cause VC loss. However, the lower content of VC of samples freeze-dried at 30 °C was even lower when higher pressure was applied (Figure 5). This means that for a long expected process time, oxygen presence should be maximally avoided.

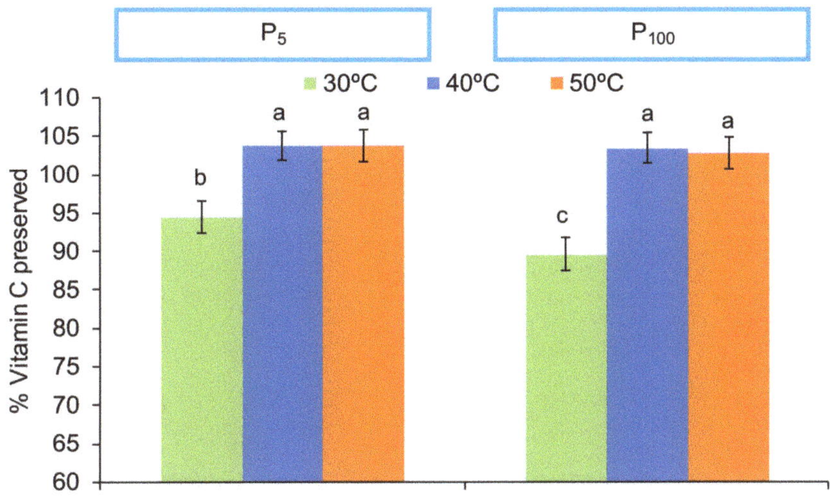

Figure 5. Percentage (%) of preserved vitamin C (mean and Tukey's HSD) according to the interaction between shelf temperature (30, 40 or 50 °C) and pressure (P_5: 5 Pa and P_{100}: 100 Pa) factors. Different letters indicate different homogeneous groups for the temperature*pressure interaction ($p < 0.05$). Data of both freezing rates are considered in the mean values.

Certain carotenoids are highly coloured compounds that also exhibit provitamin A activity. BC has the highest vitamin A activity [43]. The PLS-R (Figure 1) underlined, in particular, the BC preservation by low pressure, as it is projected on the positive side of axis 1 and highly correlated with vector P_5. In this case, the ANOVA indicated a significant effect of the three factors considered ($p < 0.05$, Figure 6), the higher the pressure, the higher the temperature, and the slower the freezing rate, the greater the loss of BC. Nevertheless, the F-Values were 88, 6 and 6 for pressure, shelf temperature and freezing rate, respectively. Again, a low F-Value in the ANOVA is correlated with no significant effect detected by the PLS-R analysis. However, the ANOVA also revealed a significant interaction between the pressure and both the shelf temperature and the freezing rate ($p < 0.05$). Figure 6 shows the interaction of pressure and shelf temperature for each FR.

According to these interactions, the pressure effect is no longer significant at 50 °C, when a significant part of BC has already been degraded by the effect of the shelf temperature. In addition, the effect of the shelf temperature was not significant at higher pressure. Furthermore, when almost no oxygen is present (P_5), BC is conserved quite well, regardless of freezing rate. It is in the greater presence of oxygen (P_{100}) when the effect of the freezing rate is significant in relation to the better preservation of BC at the FR-F. From the ANOVA results, the PLS-R analysis can be nuanced in the sense that the most recommendable way to keep the maximum carotenoids present in the orange puree during freeze-drying is when the drying stage is carried out at the lowest pressure studied and heating the freeze-dryer shelves to 30 or 40 °C, without the freezing-rate being relevant in this case.

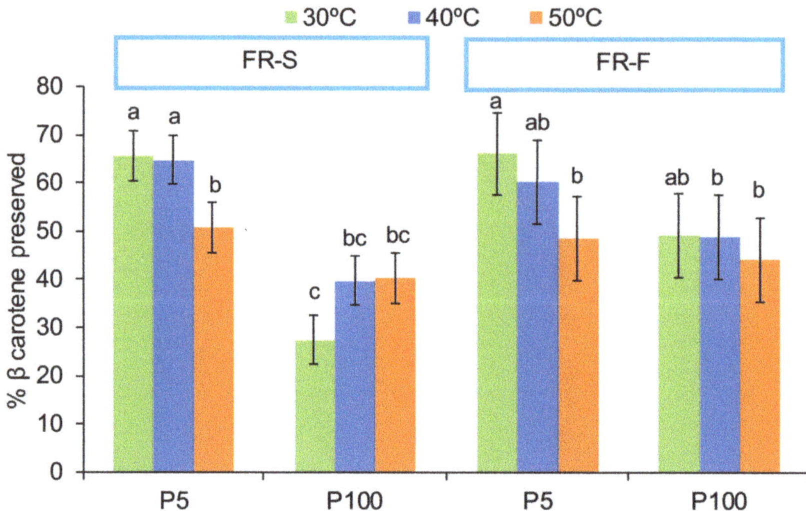

Figure 6. Percentage (%) of preserved β-carotene (mean and Tukey's HSD) according to the interaction between shelf temperature (30, 40 or 50 °C) and pressure (P_5: 5 Pa and P_{100}: 100 Pa) factors for each freezing-rate (FR-S and FR-F: slow and fast freezing rates, respectively). Different letters indicate different homogeneous groups for the temperature*pressure interaction for both freezing-rate independently ($p < 0.05$).

As regards antioxidant activity, values of DPPH between 86.5 ± 1.8% and 94.3 ± 1.5% were observed. The PLS-R revealed that the vector DPPH is projected on the lower right corner of the graph, which means a moderate positive correlation with both the lowest pressure and the highest temperature. According to the ANOVA analysis, no significant effect of pressure was detected ($p > 0.05$). Although freeze-drying carried out at 30 °C leads to samples with lower DPPH than those processed at 40 °C or 50 °C ($p < 0.05$), once again, with a low F-Value (9.51) for shelf temperature factor. On the other hand, FRAP values between 91 ± 2% and 103 ± 6% of preservation for all the conditions studied were analysed. Neither the PLS-R nor the ANOVA analysis showed a significant effect of any of the freeze-drying process variables on FRAP ($p > 0.05$).

From the Pearson correlation, only a significant and positive correlation (0.5774, $p < 0.05$) was obtained between values of DPPH and vitamin C. Despite AOA being correlated in a positive way with the total phenolic, vitamin C content and carotenoids, it has been suggested that VC contributes to antioxidant capacity more than others antioxidant constituents, such as phenols or carotenoid in fruits with high VC content [41,44]. This can also be observed on the PLS-R projection as VC, DPPH and FRAP are projected on the same direction.

4. Conclusions

In conclusion, the optimum freeze-drying conditions for preserving the nutrients considered in this study and with interesting structural properties of the obtained product, as to be perceived as crunchy by the consumers, are low pressure (5 Pa) and high shelf temperature (50 °C). These conditions also promote freeze-dried puree with a clear, yellowish and less saturated colour. The fact that a lower degradation of nutrients was observed at higher temperatures may be explained by the great reduction (75%) of the duration of freeze-drying process at 50 °C, and the mild temperatures used. The shorter exposure of nutrients to a minimal presence of oxygen in a high porous matrix is less favourable to oxidation/degradation reactions and contributes to the preservation of nutrients. As regards to the statistical analysis of the data obtained in this study, PLS-R projection may be recommended against

ANOVA as an easier tool to detect the most important factors and interactions to be considered for freeze-drying process optimization. ANOVA allows a more precise analysis, though less practical.

Supplementary Materials: The following are available online at http://www.mdpi.com/2304-8158/9/1/32/s1, Table S1: Values (mean ± SD) of the different physicochemical properties evaluated. All acronyms used are described in the main text. Table S2: Percentage (mean ± SD) of the bioactive compounds preserved in the FDP for each condition evaluated. All acronyms used are described in the main text. Table S3: Values (mean and Tukey'HSD classification) of the different physicochemical properties and bioactive compounds evaluated for: each individual factor (a–c), the interaction between two different factors (d–f) and for the interaction between the three factors studied (g). All acronyms used are described in the main text. Table S4: Variable importance in the PLS-R Projection (VIP). VIP > 1 are considered as the most important variables for the model. All acronyms used are described in the main text.

Author Contributions: Conceptualization: M.A.S.-E., M.d.M.C. and N.M.-N.; Data curation: M.A.S.-E. and M.d.M.C.; Formal analysis: M.A.S.-E., C.A. and N.M.-N.; Funding acquisition: M.A.S.-E., M.d.M.C. and N.M.-N.; Investigation: M.A.S.-E., M.d.M.C. and N.M.-N.; Methodology: M.A.S.-E., C.A., M.d.M.C. and N.M.-N.; Project Administration: M.d.M.C. and N.M.-N.; Supervision, N.M.-N.; Writing-original draft: M.A.S.-E., C.A., and N.M.-N.; Writing-review and editing: M.A.S.-E., C.A., T.F., and N.M.-N. All authors have read and agreed to the published version of the manuscript.

Funding: This research was funded by Ministerio de Economía y Competitividad of Spain [AGL 2017–89251] and by Ministerio de Educación of Spain [FPU14/02633].

Acknowledgments: The authors thank the Ministerio de Economía y Competitividad of Spain for the financial support given through the Project AGL 2017–89251 and the Ministerio de Educación of Spain for the FPU grant (FPU14/02633) awarded to Ms. Andrea Silva.

Conflicts of Interest: The authors declare no conflict of interest.

References

1. Poulose, S.M.; Harris, E.D.; Patil, B.S. Citrus Limonoids Induce Apoptosis in Human Neuroblastoma Cells and Have Radical Scavenging Activity. *J. Nutr.* **2005**, *135*, 870–877. [CrossRef] [PubMed]

2. Kader, A.A. Flavor Quality of Fruits and Vegetables. *J. Sci. Food Agric.* **2008**, *88*, 1863–1868. [CrossRef]

3. De Ancos, B.; Gónzalez, E.M.; Cano, M.P. Ellagic Acid, Vitamin C, and Total Phenolic Contents and Radical Scavenging Capacity Affected by Freezing and Frozen Storage in Raspberry Fruit. *J. Agric. Food Chem.* **2000**, *48*, 4565–4570. [CrossRef] [PubMed]

4. Aschoff, J.K.; Kaufmann, S.; Kalkan, O.; Neidhart, S.; Carle, R.; Schweiggert, R.M. In Vitro Bioaccessibility of Carotenoids, Flavonoids, and Vitamin C from Differently Processed Oranges and Orange Juices [*Citrus sinensis* (L.) Osbeck]. *J. Agric. Food Chem.* **2015**, *63*, 578–587. [CrossRef] [PubMed]

5. Gökmen, V.; Kahraman, N.; Demir, N.; Acar, J. Enzymatically Validated Liquid Chromatographic Method for the Determination of Ascorbic and Dehydroascorbic Acids in Fruit and Vegetables. *J. Chromatogr. A* **2000**, *881*, 309–316. [CrossRef]

6. Peterson, J.; Dwyer, J.; Beecher, G.; Bhagwat, S.A.; Gebhhardt, S.E.; Haytowitz, D.B.; Holden, J.M. Flavanones in Oranges, Tangerines (Mandarins), Tangors, and Tangelos: A Compilation and Review of the Data from the Analytical Literature. *J. Food Compos. Anal.* **2006**, *19*, 66–73. [CrossRef]

7. Ramful, D.; Tarnus, E.; Aruoma, O.I.; Bourdon, E.; Bahorun, T. Polyophenol Composition, Vitamin C Content and Antioxidant Capacity of Mauritian Citrus Fruit Pulps. *Food Res. Int.* **2011**, *44*, 2088–2099. [CrossRef]

8. Goulas, V.; Manganaris, G.A. Exploring the Phytochemical Content and the Antioxidant Potential of Citrus Fruits Grown in Cyprus. *Food Chem.* **2012**, *131*, 39–47. [CrossRef]

9. Moreiras, O.; Carbajal, A.; Cabrera, L.; Cuadrado, C. *Tablas de Composición de los Alimentos*; Pirámide Editions: Madrid, Spain, 2018; pp. 52–53.

10. Institute of Medicine. *Dietary Reference Intakes for Vitamin C, Vitamin E, Selenium, and Carotenoids*; The National Academies Press: Washington, DC, USA, 2000; p. 507.

11. Fazaeli, M.; Emam-Djomeh, Z.; Kalbasi-Ashtari, A.; Omid, M. Effect of Process Conditions and Carrier Concentration for Improving Drying Yield and other Quality Attributes of Spray Dried Black Mulberry (Morus nigra) Juice. *Int. J. Food Eng.* **2012**, *8*, 1–20. [CrossRef]

12. Barbosa-Canovas, G.; Ortega-Rivas, E.; Juliano, P.; Yan, H. *Food Powders: Physical Properties, Processing and Functionality*; Kluwer Academic/Plenum Publisher: New York, NY, USA, 2005; pp. 271–304.

13. Karam, M.C.; Petit, J.; Zimmer, D.; Baudelaire, E. Effects of Drying and Grinding in Production of Fruit and Vegetable Powders. A Review. *J. Food Eng.* **2016**, *188*, 32–49. [CrossRef]

14. Roos, Y.H. *Phase Transitions in Food*; Academic Press: San Diego, CA, USA, 1995; pp. 19–47.

15. Silva, M.A.; Sobral, P.J.A.; Kieckbusch, T.G. State Diagrams of Freeze-Dried Camu-Camu (*Myrciaria dubia* (HBK) Mc Vaugh) Pulp with and without Maltodextrin Addition. *J. Food Eng.* **2006**, *77*, 426–432. [CrossRef]

16. Conceição, M.C.; Fernandes, T.N.; de Resende, J.V. Stability and Microstructure of Freeze-Dried Guava Pulp (*Psidium guajava*, L.) with Added Sucrose and Pectin. *J. Food Sci. Technol.* **2016**, *53*, 2654–2663. [CrossRef] [PubMed]

17. Telis, V.R.N.; Martínez-Navarrete, N. Biopolymers Used as Drying Aids in Spray- Drying and Freeze-Drying of Fruit Juices and Pulps. In *Biopolymer Engineering in Food Processing*; Telis, V.N.N., Ed.; CRC Press: Boca Raton, FL, USA, 2012; pp. 279–325.

18. Martínez-Navarrete, N.; Salvador, A.; Oliva, C.; Camacho, M.M. Influence of Biopolymers and Freeze-Drying Shelf Temperature on the Quality of a Mandarin Snack. *LWT* **2019**, *99*, 57–61. [CrossRef]

19. Genin, N.; Rene, F. Influence of Freezing Rate and the Ripeness State of Fresh Courgette on the Quality of Freeze-Dried Products and Freeze-Drying Time. *J. Food Eng.* **1996**, *29*, 201–209. [CrossRef]

20. Karel, M.; Flink, J.M. Influence of Frozen State Reactions on Freeze-Dried Foods. *J. Agric. Food Chem.* **1973**, *21*, 16–21. [CrossRef]

21. Ceballos, A.M.; Giraldo, G.I.; Orrego, C.E. Effect of Freezing Rate on Quality Parameters of Freeze-Dried Soursop Fruit Pulp. *J. Food Eng.* **2012**, *111*, 360–365. [CrossRef]

22. Hammami, C.; René, F. Determination of Freeze-Drying Process Variables for Strawberries. *J. Food Eng.* **1997**, *32*, 133–154. [CrossRef]

23. Lombraña, J.I. The Influence of Pressure and Temperature on Freeze-Drying in an Adsorbent Medium and Establishment of Drying Strategies. *Food Res. Int.* **1997**, *30*, 213–222. [CrossRef]

24. Ratti, C. Hot Air and Freeze-Drying of High-Value Foods: A Review. *J. Food Eng.* **2001**, *49*, 311–319. [CrossRef]

25. Egas-Astudillo, L.A.; Silva, A.; Uscanga, M.; Martínez-Navarrete, N.; Camacho, M.M. Impact of Shelf Temperature on Freeze-Drying Process and Porosity Development. In Proceedings of the 21st International Drying Symposium (IDS' 2018), Valencia, Spain, 11–14 September 2018; pp. 843–850.

26. Agudelo, C.; Igual, M.M.; Camacho, M.M.; Martínez-Navarrete, N. Effect of Process Technology on the Nutritional, Functional, and Physical Quality of Grapefruit Powder. *Food Sci. Technol. Int.* **2016**, *23*, 61–74. [CrossRef]

27. Association of Official Analytical Chemists (AOAC). *Official Methods of Analysis*; Association of Official Analytical Chemists: Arlington, VA, USA, 1990; pp. 931–935.

28. Contreras, C.; Martín-Esparza, M.E.; Chiralt, A.; Martínez-Navarrete, N. Influence of Microwave Application on Convective Drying: Effects on Drying Kinetics, and Optical and Mechanical Properties of Apple and Strawberry. *J. Food Eng.* **2008**, *88*, 55–64. [CrossRef]

29. Hutchings, J.B. *Food Colour and Appearance*; Blackie Academic & Professional: Glasglow, UK, 1999; pp. 199–235.

30. Okos, M.R. *Physical and Chemical Properties of Food*; American Society of Agricultural Engineers: Saint Joseph, MI, USA, 1986; pp. 35–77.

31. Tomás-Barberán, F.A.; Gil, M.; Cremin, P.; Waterhouse, A.L.; Hess-Pierce, B.; Kader, A.A. HPLC-DAD-ESIMS Analysis of Phenolic Compounds in Nectarines, Peaches, and Plums. *J. Agric. Food Chem.* **2001**, *49*, 4748–4760. [CrossRef] [PubMed]

32. Benzie, I.F.F.; Strain, J.J. Ferric Reducing/Antioxidant Power Assay: Direct Measure of Total Antioxidant Activity of Biological Fluids and Modified Version for Simultaneous Measurement of Total Antioxidant Power and Ascorbic Acid Concentration. *Methods Enzymol.* **1999**, *299*, 15–27. [CrossRef] [PubMed]

33. Selvendran, R.R.; Ryden, P. Methods in Plant Biochemistry. In *Carbohydrates*; Dey, P.M., Ed.; Academic Press: London, UK, 1990; p. 549.

34. Brand-Williams, W.; Cuvelier, M.E.; Berset, C. Use of a Free Radical Method to Evaluate Antioxidant Activity. *LWT Food Sci. Technol.* **1995**, *28*, 25–30. [CrossRef]

35. Sánchez-Mata, M.C.; Cámara-Hurtado, M.; Díez-Marqués, C.; Torija-Isasa, M.E. Comparison of High-Performance Liquid Chromatography and Spectrofluorimetry for Vitamin C Analysis of Green Beans (*Phaseolus vulgaris*, L.). *Eur. Food Res. Technol.* **2000**, *210*, 220–225. [CrossRef]

36. Sánchez-Moreno, C.; Plaza, L.; De Ancos, B.; Cano, M.P. Quantitative Bioactive Compounds Assessment and Their Relative Contribution to the Antioxidant Capacity of Commercial Orange Juices. *J. Sci. Food Agric.* **2003**, *83*, 430–439. [CrossRef]

37. Xu, B.J.; Chang, S.K.C. A Comparative Study on Phenolic Profiles and Antioxidant Activities of Legumes as Affected by Extraction Solvents. *J. Food Sci.* **2007**, *72*, 159–166. [CrossRef]

38. Olives, A.I.; Cámara, M.; Sánchez, M.C.; Fernández, V.; López, M. Application of a UV-Vis Detection-HPLC Method for a Rapid Determination of Lycopene and β-Carotene in Vegetables. *Food Chem.* **2006**, *95*, 328–336. [CrossRef]

39. Bodart, M.; de Peñaranda, R.; Deneyer, A.; Flamant, G. Photometry and Colorimetry Characterisation of Materials in Daylighting Evaluation Tools. *Build. Environ.* **2008**, *43*, 2046–2058. [CrossRef]

40. Obied, H.K.; Bedgood, D.R.; Prenzler, P.D.; Robards, K. Effect of Processing Conditions, Prestorage Treatment, and Storage Conditions on the Phenol Content and Antioxidant Activity of Olive Mill Waste. *J. Agric. Food Chem.* **2008**, *56*, 3925–3932. [CrossRef]

41. Igual, M.; García-Martínez, E.; Camacho, M.M.; Martínez-Navarrete, N. Effect of Thermal Treatment and Storage on the Stability of Organic Acids and the Functional Value of Grapefruit Juice. *Food Chem.* **2010**, *118*, 291–299. [CrossRef]

42. Lin, T.M.; Durance, T.D.; Scaman, C.H. Characterization of Vacuum Microwave, Air and Freeze Dried Carrot Slices. *Food Res. Int.* **1998**, *31*, 111–117. [CrossRef]

43. Fisher, J.F.; Rouseff, R.L. Solid-Phase Extraction and HPLC Determination of β-Cryptoxanthin and α- and β-Carotene in Orange Juice. *J. Agric. Food Chem.* **1986**, *34*, 985–989. [CrossRef]

44. Tavarini, S.; Degl'Innocenti, E.; Remorini, D.; Massai, R.; Guidi, L. Antioxidant Capacity, Ascorbic Acid, Total Phenols and Carotenoids Changes during Harvest and after Storage of Hayward Kiwifruit. *Food Chem.* **2008**, *107*, 282–288. [CrossRef]

Article

Freeze-Drying of Blueberries: Effects of Carbon Dioxide (CO$_2$) Laser Perforation as Skin Pretreatment to Improve Mass Transfer, Primary Drying Time, and Quality

Pablo Munzenmayer [1], Jaime Ulloa [1], Marlene Pinto [1], Cristian Ramirez [1,2], Pedro Valencia [1,2], Ricardo Simpson [1,2] and Sergio Almonacid [1,2,*]

[1] Departamento de Ingeniería Química y Ambiental, Universidad Técnica Federico Santa María, 2390123 Valparaíso, Chile; pablo.munzenmayer@alumnos.usm.cl (P.M.); jaime.ulloam@alumnos.usm.cl (J.U.); marlene.pinto@usm.cl (M.P.); cristian.ramirez@usm.cl (C.R.); pedro.valencia@usm.cl (P.V.); ricardo.simpson@usm.cl (R.S.)

[2] Centro Regional de Estudios en Alimentos y Salud (CREAS) Conicyt-Regional R06I1004, 31000000 Valparaíso, Chile

* Correspondence: sergio.almonacid@usm.cl; Tel.: +56-32-2654559; Fax: +56-32-2654258

Received: 25 November 2019; Accepted: 25 January 2020; Published: 18 February 2020

Abstract: Freeze-dried berry fruits are generally consumed as they are, whole and without peeling or cutting, as the conservation of their original shape and appearance is often desired for the final product. However, usually, berries are naturally wrapped by an outer skin that imparts a barrier to vapor flow during freeze-drying, causing berry busting. Photo-sequence, experimental, and theoretical methodologies were applied to evaluate the application of CO$_2$ laser microperforations to blueberry skin. Under the same set of freeze-drying conditions, blueberries with and without perforations were processed. The results showed that the primary drying time was significantly reduced from 17 ± 0.9 h for nontreated berries to 13 ± 2.0 h when nine microperforations per berry fruit were made. Concomitantly, the quality was also significantly improved, as the percentage of nonbusted blueberries at the end of the process increased from an average of 47% to 86%. From a phenomenological perspective, the analysis of the mass transfer resistance of nontreated fruits, in agreement with reported studies, showed a Type II curvature, with a sharp decrease at low time, followed by a linear increase. In contrast, blueberries with nine perforations depicted a Type III regime, with a saturation curvature toward the time axis. It was demonstrated that CO$_2$-laser microperforation has high potential as a skin pretreatment for the freeze-drying of blueberries.

Keywords: blueberry freeze-drying; berry-busting; skin perforation; primary drying time; quality

---

1. Introduction

Freeze-drying (FD) or lyophilization is a dehydration method that has been proven to produce high-quality fruit and vegetable products, as compared with other drying processes. Since it is carried out at a low temperature (typically a maximum of 38 to 54 °C shelf temperature), product damage is minimized. Thus, it is used for delicate, high-sensitive, and high-value products to maintain their color, flavor, shape, and nutritional characteristics [1–3]. Examples of these kinds of foods are berry fruits such as blueberries, strawberries, maquiberries, and cranberries, among others. Freeze-dried berry fruits are generally consumed as they are, whole and without peeling or cutting, as conservation of their original shape and appearance is often desired for the final product [4]. However, most of these berries are naturally wrapped by a waxy outer skin that imparts barrier to water vapor flow during FD, as happens in all drying processes. This mass transfer limitation is often the phenomenological factor that controls the processing time, energy consumption, and concomitant final quality. In some processes, such as FD, where fruits are exposed to high-energy inputs and/or a high-vacuum environment, a skin rupture or a busting process is frequently observed due to pressure lift just under the skin. As a result, besides the long processing time, many freeze-dried berries, while maintaining most of their original nutritional quality, cannot keep their original shape and appearance, diminishing their organoleptic characteristics significantly [5–8].

Several skin pretreatment methods have been tested to facilitate water vapor transfer in drying processes, most of them focused on reducing the processing time of Osmotic Dehydration (OD). Physical and chemical methods, such as maceration of the outer skin by knife blade, chemical additives, or needle perforation, have been reported [1,9–14]. Due to the presence of undesirable compounds (NaOH, HCl, ethyl oleate), the generation of waste material, and the low quality found in the final product, chemical treatments are not a real alternative to weaken the berry skin. For example, Grabowski and Marcotte [13], reported that chemical skin pretreatment in OD of cranberries resulted in the lowest value of taste acceptability. Different results were reported by Ketata et al. [15]: when liquid nitrogen pretreatment was applied to the OD of blueberries, the dewaxing of the fruit skin allowed a reduction in the drying time of 45% to 65%; however, this was accompanied by a significant loss of total phenolics. Mechanically cutting fruits is another alternative that has been tested in blueberries and tomatoes. However, due to the softness of these fruits, many problems have been observed, including significant damage to the final product, loss of nutritious fluids, and not having the whole fruit as the final product [1]. Alternatively, skin needle perforation has been tested on cranberries and cherry tomatoes. It has been reported that to obtain a significant mass transfer enhancement in OD, 20% to 30% of the total surface area of cranberries should be perforated [13]. Azoubel and Murr [10] perforated cherry tomatoes with 1 mm diameter needles with a pin hole density of 16 holes/cm^2 prior to OD and air drying, attaining a significant time reduction when more than 80 holes/cm^2 were used.

Scharschmidt and Kenyon [16] reported the results related to skin pretreatment in FD processing where blueberries were perforated by needles to a density of 2–3 punctures per berry. The main results were that the berries kept their original spherical shape without physical changes to their outer integument, which was declared to be a problem in nonperforated blueberries. In addition, the FD time was two-thirds or less than that of the control process without puncturing. However, no information was given on processing conditions (temperature and pressure), and also the needle diameter was not specified. Thus, even with some time-limited information, physical skin perforation of the whole berry fruit (i.e., blueberry) seems to be a suitable alternative to more efficiently carry out mass transfer processes that reduce the processing time, cost, and energy requirement and avoid product explosion or bust, so that the fruit can retain its basic shape.

Collaterally, puncturing has been tested by utilizing carbon dioxide (CO_2) laser beam technology to carry out surface microperforations. This has important advantages, and it is already used in many fields such as medicine, cosmetics, and marking industries. Due to its superior accuracy, environmental cleanliness, and safety, applications can be carried out at precise locations, in multiple surfaces and arrangements, and within a size range of 50 to 300 μm [17,18]. Also, it is a noncontact technology that significantly mitigates the chance of physical and microbiological contamination of materials that are typically associated with traditional cutting or contact devices or fluids [19,20]. Fujimaru et al. [9] reported the application of CO_2 laser microperforation of blueberry skin as a pretreatment in OD. The study demonstrated that CO_2 laser microperforation can be a viable skin pretreatment which offers notable improvements in water removal; it almost doubled the moisture loss in the first 24 h of OD, from 7% (control) to 11% (2.5 mm square grid perforation), which was even better than the 9% water loss for blueberries that were mechanically cut. This already-proven technology, which the food industry has not completely incorporated, is an attractive alternative that can be applied to berry FD in order to overcome the problems associated with its outer skin.

From a theoretical point of view, the busting process is a consequence of a high initial mass transfer resistance due to the presence of the skin, with this being probably the most significant variable governing water vapor flow during FD and its implications. When the resistance to mass transfer needs to be estimated, the Manometric Temperature Measurement (MTM) methodology has been proven to be an effective tool. Moreover, Pikal et al. [21], concluded that a high resistance to mass transfer at the beginning of the sublimation process is due to the presence of a surface barrier resulting from a structure different from that of the dried layer, a conclusion that has been supported by scanning electron microscopy. This observation may be applied to berry FD, but it needs to be tested.

Thus, the objectives of this study were as follows: (1) to describe the busting process of blueberry freeze-drying through experimental and phenomenological approaches, and (2) to evaluate the use of CO_2 laser microperforation technology to reduce the processing time as well as product explosion and its associated consequences.

1.1. Theoretical Background

1.1.1. The Freeze-Drying Processes

Similar to other dehydration processes, FD includes simultaneous heat and mass transfer. A simple but conceptual explanation of the phenomenological processes taking place during a typical like-berry-fruit FD operation is described in Figure 1.

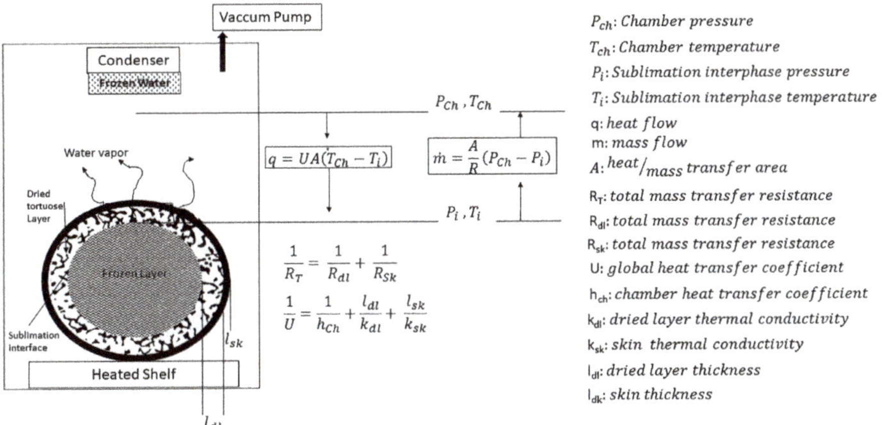

Figure 1. Like-berry-fruit freeze-drying system and its basic phenomenological description.

A product, initially completely frozen, is placed into a freeze dryer chamber. Then, the chamber pressure is lowered under the water triple point, and heat is supplied by mean of a heated shelf and/or the surrounding air temperature (primary drying or sublimation). As heat flows through the dried layer through heat transfer, sublimation latent heat is delivered, and a sublimation interface recedes leaving, resulting in a thicker porous layer of dried material, which acts as a resistance to heat transfer, as well as water vapor flow, toward the product surface through mass transfer, and finally, to the condenser trap, where water is separated before gases are expelled by the vacuum pump. Both heat and the mass transfer rate are the essential causes that make the FD process slow down. In FD, if heat supplied to the sublimation interface is equal to the latent heat associated with the sublimation rate of ice (m × ΔH_S = q), the saturated pressure of sublimation (P_i) and its corresponding temperature (T_i) become stable, and the sublimation proceeds normally. However, if the heat supplied to the sublimation interface is not enough, the sublimation rate will drop; conversely, if the total resistance to vapor diffusion is too large (R_T), then P_i and T_i may rise. As a result, frozen water will melt and collapse, and/or bust may take place [20,21]. During berry-fruit FD processing, in addition to the dry-layer resistance (R_{dl}), the skin operates as an important barrier (R_{sk}) that further extends the processing times. Additionally, the low skin permeability can generate explosion or bust problems, as vapor cannot escape at the needed rate so as not to provoke a pressure lift just under the skin [5].

1.1.2. Mass Transfer Resistance

The estimation of R_T through experimental and theoretical approaches would explain and describe the phenomenological process associated with the berry busting process. The interaction effects between the blueberry fruit characteristics, such as mass transfer area (A) the product mass transfer resistance (R_T), and the freeze-drying processing conditions, have been studied by using the semiempirical MTM procedure, which was originally developed to assess the temperature of the sublimation interface within the product and the dried-layer mass transfer resistance. The method has a large body of literature information [22–27]. The principle of MTM is based on the flow of water vapor from the product chamber to the condenser being momentarily interrupted during the primary drying time. The MTM is based on a rapid increase in the chamber pressure by closing the freeze-dryer butterfly valve (Figure 2). During this perturbation process, the chamber pressure will rapidly increase due to the continued sublimation of ice. Since the composition of the vapor phase in the chamber is nearly all water vapor, sublimation will stop when the chamber pressure reaches the vapor pressure of ice at the sublimation interface (diving force $\Delta P = 0$); consequently, the pressure rise will cease. The dynamics of this pressure rise process can be theoretically described by equation (1), which is derived from the phenomenological heat and mass balance over a system, defined as the void volume of the product chamber [28]:

$$P(t) = P_i - (P_i - P_0)Exp\left(\frac{-NART_s}{M_{H_2O}VR_T}t\right) + 0.0465P_i\Delta T\left[1 - 0.811Exp\left(\frac{-0.114}{L}t\right)\right] + Xt \qquad (1)$$

where $P(t)$ is the system pressure (void volume of the freeze-drying chamber), P_i is the interphase pressure of the product, P_0 is the initial pressure of the system, N is the number of fruit units, M_{H2O} is the water molecular weight; V is the void volume of the freeze drier system, R is the universal gas constant, T_S is the shelf temperature, A is the mass transfer area per fruit unit, R_T is the total area normalized product resistance to mass transfer, ΔT is the temperature gradient across the frozen layer—normally fixed at 1 K, L' is the geometric product characteristic—ice thickness, and X is the linear term in equation (1) associated with external air infiltration. Tang et al. [28] generated simulations of pressure rise curves to analyze the relative contribution to the chamber pressure rise ($P(t)$), of the many terms in Equation (1), concluding that a resistance-dominated period, which takes place in the first 5 to 10 s of the pressure rise, is mainly expressed by the first exponential term, so the others terms are nearly neglected as part of the summation.

2. Materials and Methods

2.1. Materials

The experiments were carried out with fresh blueberries of the Huertos de Chile brand (variety *Vaccinium corymbosum*), which were bought at a local market. Blueberries with a diameter of 14–15 mm were selected.

2.2. Blueberry Characterization and Freezing

The blueberry fruits used as raw material to carry out a set of experimental treatments were characterized to ensure a homogeneous quality of said material. Total Soluble Solids (TSS), or °Brix, was selected for this purpose as it is one of the main quality indicators of the blueberry fruit [29]. To determine °Brix, blueberry fruits were weighed using a RADWAG balance (AS 220/C/2, Radom, Poland) and then crushed and homogenized for 5 min with 10 mL of distilled water until a homogeneous juice was obtained. The resulting juice was filtered, and aliquots were taken for °Brix analysis with a HI 96,680 digital refractometer (Hanna instruments, Rothe Island, USA).

Blueberries to be used in the set of experimental treatments were first frozen in a convection freezer (MT05, Alaska, Minas Gerias city, Brasil) at −40 °C for 2 h to obtain a product temperature equal to or less than −35 °C.

A series of experiments, divided into two steps, was carried out in order to achieve the research objectives:

(i)   (Blueberries were freeze-dried without any treatment, where the busting process was observed through photographic monitoring. The fraction of busted blueberries was computed throughout time.

(ii)   The same freeze-drying process was applied to blueberries under different treatments:

    a.   Whole, without any treatment.

    b.   Cut in half.

    c.   With 1, 3, 6, and 9 CO_2-laser microperforations.

The Pressure Increase Test (PIT) was performed to estimate the treatment effect on primary drying time, and the MTM method was applied to evaluate the mass transfer phenomena, as affected by blueberry characteristics (skin, A, R_T) and the busting process. The final fraction of busted blueberries was also evaluated. Table 1 depicts a summary of the experimental setup, showing its two steps—visual observation/evaluation of the busting process of whole blueberries and a total of 6 treatments to evaluate the effect of CO_2-laser microperforations on primary drying time and quality.

Table 1. Description of the experimental setup.

Process/Activity	Busting Process Study	Treatment Number					
		1	2	3	4	5	6
Blueberry Characterization	•	•	•	•	•	•	•
Whole, no perforation	•	•					
Cut into half			•				
CO_2-laser 1 perforation				•			
CO_2-laser 3 perforation					•		
CO_2-laser 6 perforation						•	
CO_2-laser 9 perforation							•
Freeze-drying process: 20 °C; 0,13 mbar	•	•	•	•	•	•	•
Photo camera monitoring for visual inspection	•						
PIT test, for Primary-drying time estimation		•	•	•	•	•	•
MTM test, for skin/dry-layer Resistance estimation		•	•				•
Evaluation of busted blueberry fraction	•	•	•	•	•	•	•

2.3. Freeze-Drying Process

All experiments were carried out in a Martin Christ freeze-dryer, model Alpha 2-4 LSCplus (Martin Christ Gefriertrocknungsanlagen, Osterode, Germany), which can operate at a total vacuum pressure of up to 0.001 mbar, provided with an MKS Baratron 622 capacitance manometer (MKS Instruments) and a condenser that can operate at temperatures down to −85 °C (see Figure 2). It has three shelves of 0.021 m^2 each that can be temperature controlled with a wireless temperature monitoring system, which, in turn, allows for sample temperature monitoring through pt100 port sensors. The samples or blueberries were placed onto the three shelves, considering a total load of 60 units. In order to assess the product temperature, one blueberry fruit per shelf was inserted with a miniature pt100 sensor (PT 100 Mini-Epsilon LSC Plus, Martin Chris, ostero de am Harz, Germany). A standard freeze-drying processing condition was fixed at 0.13 mbar (13 Pa) with a condenser temperature of −85 °C and a shelf temperature of 20 °C, while the air temperature was also 20 °C. The freeze-drier system includes dedicated software (SCADA Software V1 LPC plus Martin Chris, ostero de am Harz, Germany) to set, control, and monitor the freeze-drying process at the selected processing condition. It samples data every 5 s. A butterfly valve with an approximate closing time of 0.5 s, mounted in the cylindrical duct connecting the freeze-drying chamber and the condenser chamber, can be managed by the software to perform Pressure Increase Tests (PIT) (Figure 2). The SCADA Software LPC plus software allows graphic visualization and recording of several process variables: tray temperature, blueberry temperature, condenser temperature, chamber absolute pressure (capacitive sensor), and PIT data.

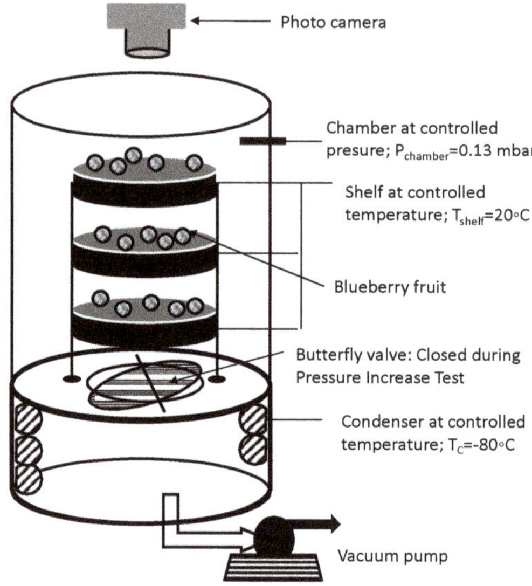

Figure 2. Freeze-drying system arrangement and its processing condition set up.

2.4. Determination of Primary Drying Time

After freezing, the FD process proceeds to primary-drying or sublimation, and finally to secondary drying or water desorption. The primary drying stage—conventionally, the most time-consuming part of the process—was investigated in the present study. In order to determine the end of the primary drying time or the transition from the primary drying stage to the final secondary-drying, which can be estimated with the automatic Pressure Increase Test (PIT), the freeze-drier software system was implemented. It works by temporarily closing the butterfly valve between the product chamber and the ice condenser (see Figure 2), allowing a pressure increase in the product chamber. If the pressure increase remains below a set limit, usually 10% whilst the valve is closed, the software program assumes that there is no further sublimation water left in the product, and the primary drying phase can be considered to be finished, and this time is recorded. Another way to estimate the end of the primary drying stage is to observe the difference between the shelf and assessed product temperature; when they equalize, no more sublimation heat is absorbed, meaning the sublimation is over.

2.5. Visual Registry of the Blueberry FD Process

A photo-sequence methodology was used to capture and visually analyze the dynamics of the busting process while blueberries are being freeze-dried. Images of the top-shelf blueberries were taken by a photographic camera (Flea®3 FL3-GE-03S2C-C Color GigE Camera, FLIR Systems, Wilson Ville, OR, USA) located at the upper part of the freeze drier system (Figure 2). A dataset of images taken at a rate of 12 photographs per min was then visually inspected to calculate the fraction of busted blueberries over time.

2.6. Estimation of the Dry-Layer Mass Transfer Resistance (R_T)

The MTM method was used to evaluate the influence of product characteristics in the FD process. The needed chamber pressure-rise as a function of time can be experimentally monitored, and the acquired data can then be regressed through Equation (1), where unknown parameters such as P_i, X, A_T (where $A_T = N \times A_{unitary}$), and R_T can be estimated. The minimum conditions on the experimental procedure have been reported in order to obtain reliable results. The data collection time could not be longer than 30 s, since a longer time would allow a considerable product temperature increase because of the closed system, and within that time, there needed to be sufficient data collection to observe the complete development of the exponential part of Equation (1), which accounts for the pressure rise controlled by the product resistance. During pressure increase runs, the dynamic pressure increase was monitored with a cDAQ module/9215 data acquisition system and LabVIEW software, manufactured by National Instruments (11500 N Mopac Expwy, Austin, Texas, United States), allowing a sampling period of as low as 10 ms. To successfully apply the MTM method, the product of the geometric characteristics and the particular void volume of the chamber system needed to be adjusted. It has been demonstrated that the computed value of the exponential expression Q (without considering t) should be equal to or higher than 0.2 (1/h) to ensure complete depiction of the exponential phase of the pressure rise in the product chamber [28]:

$$Q = \left(\frac{3.461 N A T_s}{V R_T} \right) \geq 0.2 \ (1/h) \tag{2}$$

For a given value of R_T, Q allows the minimum number of units (N, number of blueberry fruits) that must be loaded in a particular freeze-drier system (characterized by its void volume V) to be estimated to obtain a value equal to or higher than 0.2 (1/h). A total resistance value (R_T) of approximately 3 (torr h cm^2/g) has been reported to be acceptable to carry out this assessment.

2.7. CO$_2$ Laser Microperforations and Laser System Settings

As mentioned in the introduction section, CO$_2$ laser microperforation is a convenient pretreatment as it is a noncontact technology that significantly mitigates the chance of physical and microbiological contamination of materials that are typically associated with traditional cutting or contact devices or fluids. In this respect, the most-used pretreatment technology is to machine-cut the blueberry fruits into halves, which effectively avoids the busting process and reduces the primary drying time. Then, both pretreatment technologies were evaluated: CO$_2$ laser perforation at four levels of perforation—1, 3, 6, and 9 perforations per blueberry fruit—and FD blueberry cut into halves. Finally, the results were compared with each other and with those of FD whole blueberries.

A 100 W CO2-laser system (Firestar t100, Synrad Inc., Mukilteo, Wash., U.S.A.) was used to carry out the perforations of blueberries. The system was equipped with a 125 mm focusing lens (FH series Flyer, Synrad Inc.) and a computer interface with laser marking software (WinMark Pro, Synrad Inc.).

The system was operated at a continuous wavelength of 10.6 µm and a frequency of 10 kHz. Perforations were made in a square grid pattern with a density of 2.0 × 2.0 mm.

The CO2 laser was set at 120 pulses, duration of 1 millisecond, linear speed of 100 cm per second, distance of 128 mm between the laser and the surface of the blueberry, and a percentage of power that was determined experimentally by observing in a microscope (Helmut Hund GmbH, H600/12) the depth of the perforation, until finding the configuration of the laser that allowed us to perforate until 1/3, varying the power of the laser (whose maximum was 100 watts).

A unit of frozen (−35 °C) blueberry was loaded onto an aluminum tray (20 cm × 20 cm) to perform microperforations; then, blueberry samples were refrozen at −35 °C and kept until FD experiments.

3. Results and Discussion

3.1. Blueberry Characteristics

Blueberry size was selected within the diameter range of 14 to 15 mm. Assuming a spherical shape for the blueberry fruit, and, given a diameter of 14.5 mm (1.45 cm), its surface area ($A = 4\pi r^2$) results in a unit area of 6.6 cm^2.

Total Soluble Solids of the blueberries used as raw material for experimental procedures showed a characteristic quality indicator within the narrow enough range of 9.4 ± 0.7 °Brix.

3.2. Busting Process of Nontreated Whole Blueberries

The whole blueberry FD process was set as described in the methodology section.

A slide dataset that consists of images extracted from the photo sequence every 20 min was used to visually quantify the fraction of busted blueberries over time; this evolution is depicted in Figure 3. The busted blueberries are identified by red circles, reflecting the progression of the busting process, which starts with zero busted at 20 min and describes a saturation curve, achieving a maximum of 26 busted blueberries at 180 min.

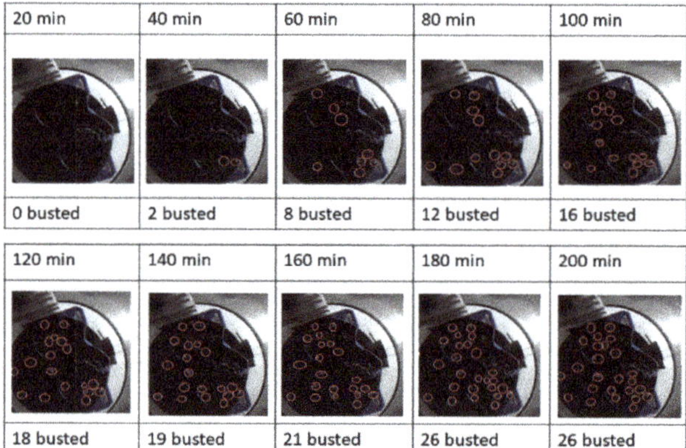

Figure 3. Visual evaluation of the busting process. Inspection of photographs taken at different times during the freeze-drying of whole nontreated blueberries.

Note that the dynamic of the busted blueberry percentage increased by up to 43% after 3 h of freeze-drying, and then it stayed constant (Figure 4). This means that the resistance to the mass flow of sublimated water is high at the beginning, and then it decreases as the number of busted blueberries increases. Then, a normal FD process should continue with increasing resistance as the dried layer gets thicker. The main process parameters and measurements of the whole blueberries FD, treatment 1 in Table 1, are depicted in Figure 5.

As expected, the percentage pressure increase (PIT) was low at the beginning of the process until around 2.5 h, in agreement with that shown in Figure 4. Then, it stayed approximately constant for 2.5 h, before decreasing at a moderate rate until 13 h, and finally, a very low decrease continued to get down to values under 10% after around 17 h. The product temperature also achieved the shelf temperature at approximately 15 h. Two important outcomes can be extracted from this result. First, normally, the resistance to vapor flow from the FD product is low at the beginning of the process, as no dried layer exists; however, in this case, resistance seems to have been high at the process initiation due to the high skin resistance. Then, as the busting process proceeds, the resistance goes down, before finally increasing again as the dried layer develops. Secondly, the results indicate that the primary drying time is in the range of 15 to 20 h. From the experiments carried out in triplicate, the end of the primary drying time was assessed by the product of the thermocouple response and PIT, and there was good agreement between the two techniques, with the global primary drying time being 17 ± 0.9 h with a final moisture content of 14% ± 2%.

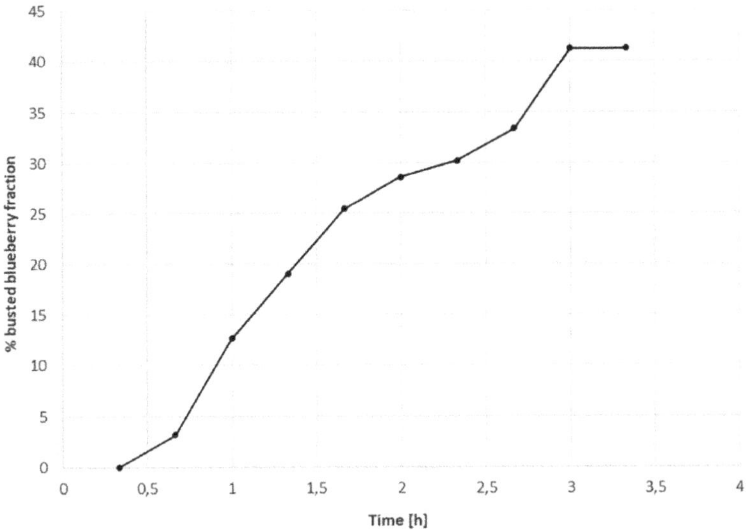

Figure 4. Busting process dynamics as evaluated from photographs taken during the freeze-drying process of whole nontreated blueberries.

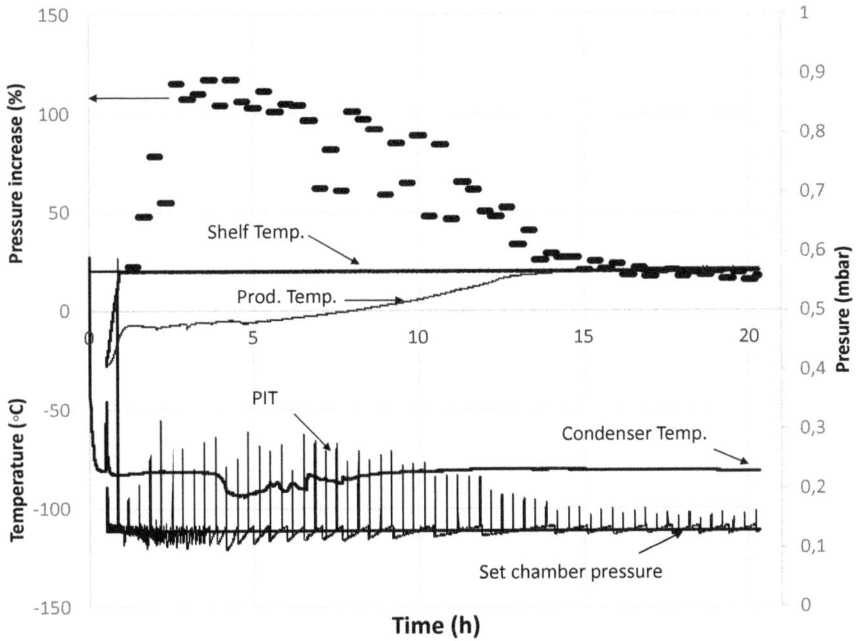

Figure 5. Main processing variables and measurements during the freeze-drying of whole nontreated blueberries, PIT.

In order to quantify the effect of the blueberry skin on mass transfer, the MTM method was applied. It has been demonstrated that the computed value of the exponential expression of Equation (1) (Q; without considering t) should be equal to, or higher than, 0.2 (1/h) (Equation(2)) to ensure complete depiction of the exponential phase of the pressure rise in the product chamber. This phase reflects the influence of the resistance (R_T) on the mass transfer phenomena. For a given value of R_T, Equation (2) allows us to estimate the minimum number of fruit units N (number of blueberry fruits) that must be loaded in a particular freeze-drier system (characterized by its void volume V). A value of R_T of approximately 3 (torr h cm^2/g) has been reported to be acceptable to carry out this assessment [28]. To satisfy the above condition, the following set of values were considered:

R/M_{H2O} = 3.461 (torr L/g-K);

N ≥ 60 units;

A = 6.6 (cm^2/unit);

V = 32.5 (L);

T_S = 293 K;

R_T = 3.0 (torr h cm^2/g).

These data were input into Equation (2), resulting in a Q value of 4.1 > 0.2.

All published studies about the application of the MTM method have used a cylindrical shape (vials in the pharmaceutical industry), where the mass transfer area (A) does not change as the FD proceeds (one-dimensional axial mass transfer). However, assuming a spherical geometry to represent the blueberry shape, area A does change over time. For this reason, in this study, the quotient (A_T/R_T), was taken as a regression variable, together with P_i and X (see Equation (1)). Experimental data of the pressure rise vs. time over 30 s were generated every 20 min. The pressure rise data were taken at a rate of 100 data/s and then averaged to generate a data set of 5 data points per second. These data were then regressed against Equation (1) to obtain (A_T/R_T). In order to show the evolution of the process, Figure 6 depicts the first 10 s of some of the pressure rise vs. time experimental data and their respective regression curves. As expected, the regressions showed neglected values for variable X (in the range of 0.001 to 0.0001), indicating a good seaming of the FD system (no air infiltration). Also, the change of P_i values was minor in comparison with the changes of (R_T/A_T), meaning that the latter is the main agent of chamber pressure rise [28].

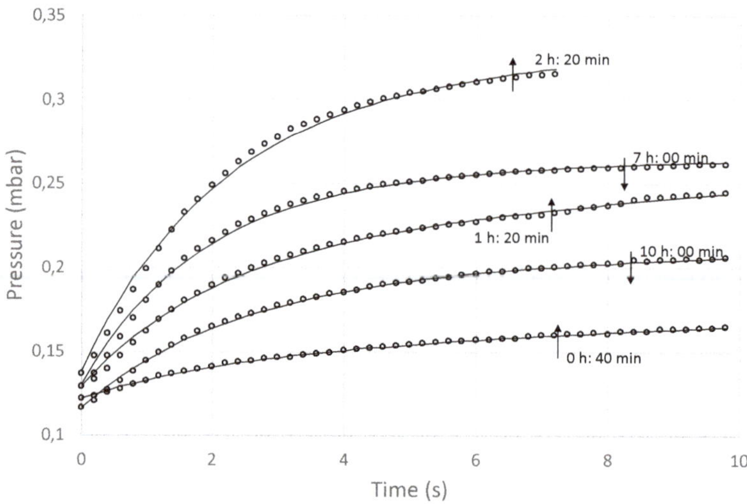

Figure 6. Fit of MTM equation to experimental pressure rise profiles for progressive times along the freeze-drying process of whole nontreated blueberries.

It can be seen that at the beginning (0 h 40 min), the pressure rise was low, and it kept increasing up to 2 h 20 min. Then, it started to decrease, approximating the pressure rise observed at the process initiation (10 h 00 min). The (A_T/R_T) values obtained by the regression analysis at different processing times, together with a tendency line, are shown in Figure 7.

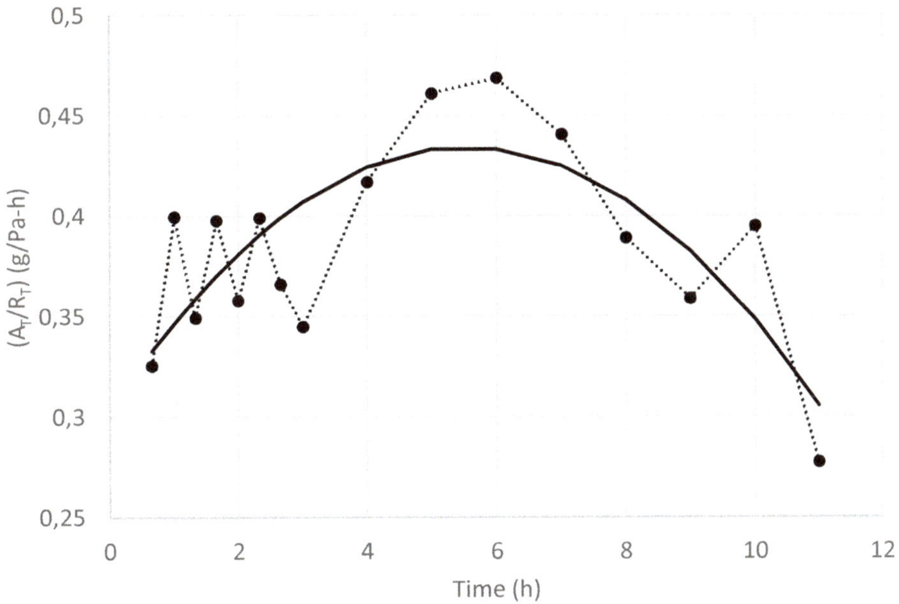

Figure 7. Values of (A_T/R_T) along the freeze-drying time (•) and its corresponding tendency line (—) of whole nontreated blueberries.

This value started at a low value, which means that a high R_T value has an important effect in the quotient (A_T/R_T), and it increased rapidly over time up to 4 h, as some of the blueberry fruits busted and R_T decreased, in spite of the A_T value that should have been decreasing during this period. This reveals that the main effect is that given by the important reduction of R_T (busting process). If an average A_T over time is assumed (for example, $A_T = 31.4$ cm^2), the resistance R_T as a function of time can be approximated, and its result is shown in Figure 8, including a tendency line, where, contrary to a normal FD process, R_T started at a high value and then decreased sharply before finally increasing again.

This observation reflects the presence of a surface barrier, which appears to be the blueberry skin, which, after being exposed to increasing vapor pressure for some time, cracks, and the effective resistance decreases. This event does not occur at the same time for all FD blueberry fruits, resulting in the sharp, but not instantaneous, R_T reduction. This is in agreement with what was observed by Lu and Pikal [30], Pikal et al. [21], and Pikal [23], where some experiments using water solutions with different solutes showed initial values of R_T significantly larger than zero, which suggests a surface barrier resulting from a different structure for the dried product near the surface, which was attributed to the formation of a high-resistance "skin" during the freezing step of the FD process, a conclusion that was supported by scanning electron microscopy. These authors, after performing various experiments with different compositions, suggested four types of resistance curves: Type I, with a linear dependence of R_T on the dried layer thickness (or concomitant elapsed time); Type II. with a sharp decrease in R_T after a short period of time, followed by a linear increase in R_T with increasing time; and Type III and IV, with a saturation curvature toward the time axis, with Type III being less severe than Type IV. The bust process described in this study is clearly a Type II resistance curve.

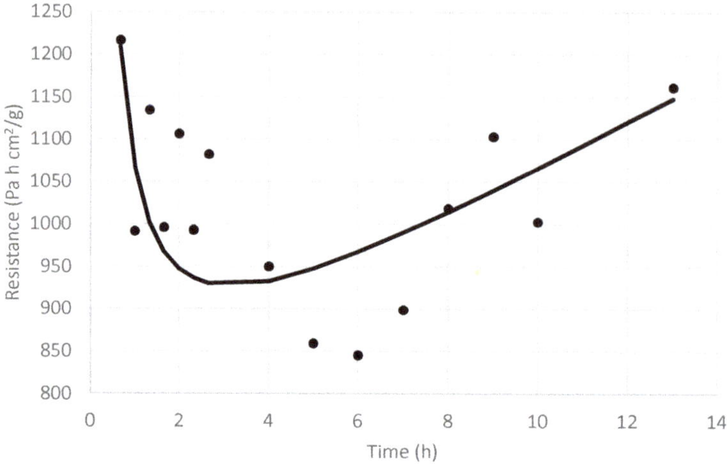

Figure 8. Values of (R_T) along the freeze-drying time (•) and its corresponding tendency line (—) of whole nontreated blueberries.

These results indicate that the reduction of the initial value of R_T, considering a pretreatment to the blueberry skin, would enhance the water vapor flow at the beginning of the process, and consequently would reduce the primary drying time and/or increase the quality yield by bringing down the fraction of busted blueberry units at the end of the process.

3.3. Evaluation of CO$_2$ Laser Microperforation and Blueberries Cut in Half

Experimental results were obtained for all five pretreated blueberries: 1, 3, 6, and 9 perforations and cut in half, treatments 1 to 6 in Table 1. Microperforations were made as described in the methodology, obtaining perforations in a square grid of 2.0 mm × 2.0 mm density, 0.5 mm diameter, and 1/3 of the blueberry diameter in depth, by utilizing 19% power (19 W). The FD system was set at the same conditions used for whole nontreated blueberries, and the same number of blueberry units were loaded in each experiment. The final moisture at the end of primary drying was also controlled and was within the range of 13% (95% CI (10%, 17%)) to 15% (95% CI (12%, 18%)), for all treatments. Figure 9 shows a blueberry with nine and three CO$_2$ laser microperforations in a square arrangement with 2.5 mm spacing. It demonstrates that the CO$_2$ laser microperforation process is minimally invasive with a negligible impact on quality characteristics.

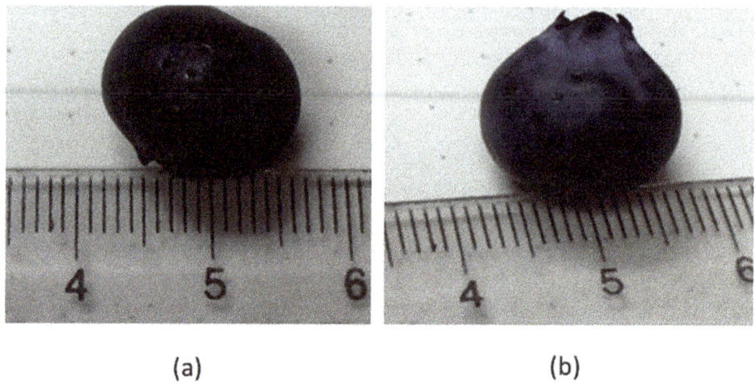

Figure 9. Appearance of blueberry fruit with nine (**a**) and three (**b**) CO$_2$ laser microperforations.

The PIT and product temperature response were used to evaluate the effect of pretreatment on the primary drying time. Each test was conducted in triplicate to determine the average and standard deviation of the primary drying time. Figure 10 depicts the PIT for the blueberries that were whole and nontreated, cut in half, and microperforated nine times. It is interesting to note that for both cut-in-half and nine-times-perforated blueberries, the initial PIT value started higher than that of the whole nontreated fruit and decreased over time. This is due to the weakened pretreated skin or the area without skin being available to sublimation.

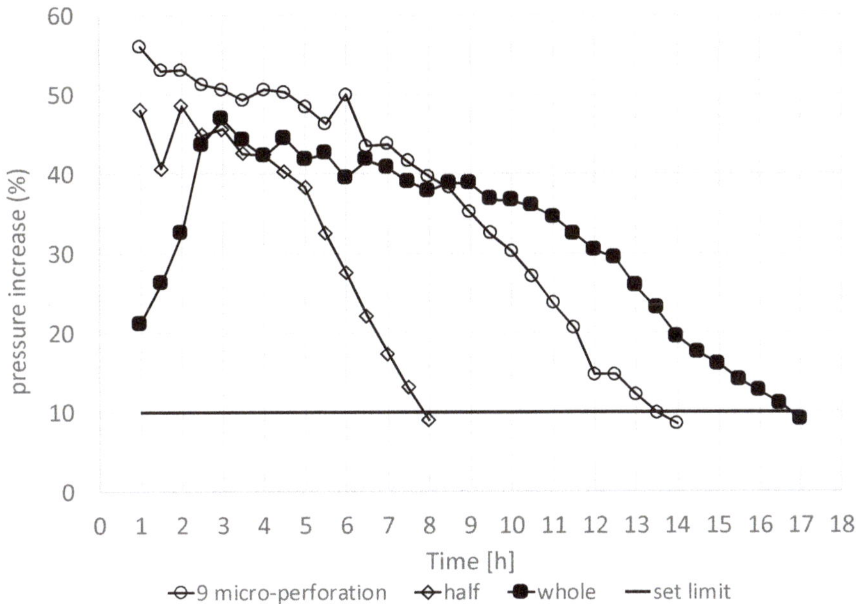

Figure 10. Pressure Increase Test (PIT) applied to blueberry freeze-drying to estimate primary drying time as affected by CO_2 laser microperforations and cut-in-half treatments.

If the pressure increase was equal to or less than 10%, then the primary drying time was considered to be over.

Table 2 shows the results of the primary drying time for the most significant treatments (treatments 1, 6, and 2 in Table 1).

Table 2. Primary drying time of blueberry freeze-drying as affected by CO_2 laser microperforations and cut-in-half treatments.

Microperforations	Time [hours]
Without microperforations	17.0 ± 0.9
Nine microperforations	13.0 ± 2.0
Cut in half	6.7 ± 1.2

It can be seen that there was an important primary drying time reduction of 60.6% when blueberries were cut in half, as compared with that of whole nontreated blueberries. However, this option is associated with the loss of the important quality characteristic of maintaining the fruits' original shape. On the other hand, and not less importantly, it can be observed that primary drying time reduced by approximately 23.5% when the blueberry skin was pretreated with nine microperforations, as compared with that of whole nontreated blueberries. This result is explained by the fact that the nine-microperforation pretreatment avoids the initial high resistance (R_T) at the beginning of the process, mitigating the busting process. This means that, in addition to primary time reduction, CO_2 laser microperforation pretreatment can improve the final product quality. It can be seen from Figure 11 that the percentage of nonbusted blueberries significantly increased as the number of perforations per fruit increased, which is statistically sustained by a positive value of the covariance (covariance = +0.634). In addition, the average value of the percentage ofnonbusted blueberries with nine microperforations was 86%, significantly higher than 47% for those with no microperforations. This is supported by their corresponding 95% CIs:

Zero microperforations: average 47%; 95% CI (33%, 61%);

Nine microperforations: average 86%; 95% CI (73%, 99%).

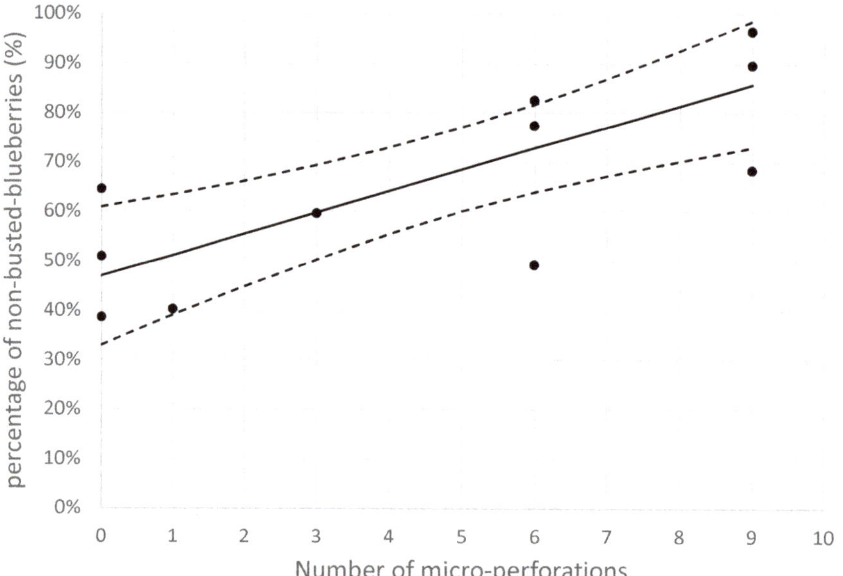

Figure 11. Experimental and statistical analysis of the nonbusted percentage (%) as affected by CO_2 laser microperforations.

Figure 12 also shows that pretreated blueberries are significantly better than those that can be purchased in a local market.

Similar to the analysis made to the whole blueberry process, pressure rise data were regressed against Equation (1) to obtain values of (A_T/R_T) to finally assess the effect of two pretreatments (nine perforations and cut in half) on R_T behavior during the FD process and compare them with that of the whole nontreated blueberry process. Figures 13 and 14 depict the resistance behavior of cut-in-half and nine-perforation treatments over time, respectively. In both cases, it can be observed that there is a sort of lag period for the R_T value, which can be attributed to the fact that the skin resistance (R_{sk}) is still significantly higher than that of the dried layer, with this lag period being longer for the nine-perforation pretreatment than that of the cut-in-half treatment (about 3 h difference). After the lag period, both of them behave like the normal FD process.

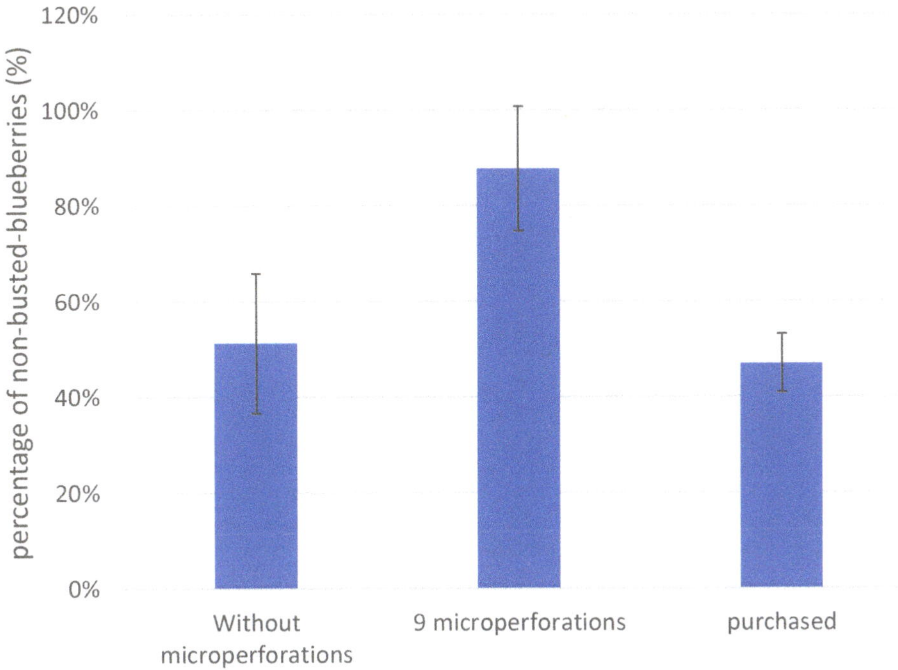

Figure 12. Differences between freeze-dried blueberries as affected by pretreatment and comparison with those that can be found in local markets.

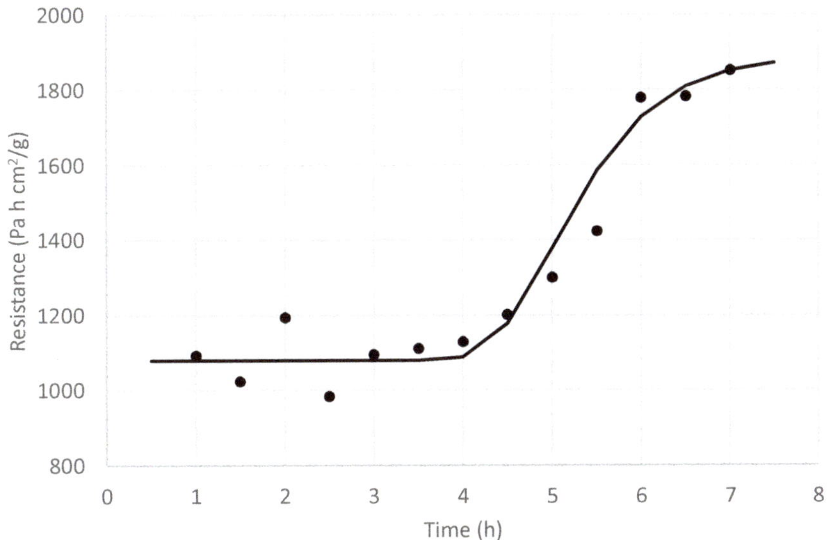

Figure 13. Values of (R_T) along the freeze-drying time (●) and its corresponding tendency line (—) of cut-in-half treated blueberries.

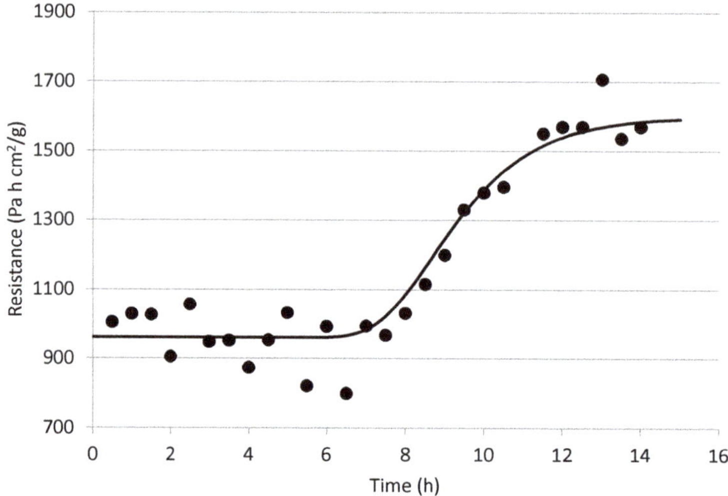

Figure 14. Values of (R_T) along the freeze-drying time (•) and its corresponding tendency line (—) of nine CO_2 laser microperforations treated blueberries.

In order to compare all three treatments—whole nontreated, nine-perforation, and cut-in-half blueberries—the tendency lines were represented in the same graph (Figure 15). This figure summarizes the phenomenological effect of the skin and pretreatments on the blueberry freeze-drying, an effect that is reflected by the R_T behavior during FD processing. According to the classification made by Pikal et al. [21], the whole nontreated blueberry FD process, as mentioned, would follow a Type II resistance curve; the cut-in-half blueberries would follow a Type IV process with a severe curvature toward the time axis; and finally, the nine-perforation blueberries would exhibit Type III behavior, with a less severe saturation curvature toward the time axis than that of Type IV. These outcomes phenomenologically describe the effect of the skin and pretreatments on the mass transfer process, which is mainly reflected by the evolution of R_T throughout the FD processing time.

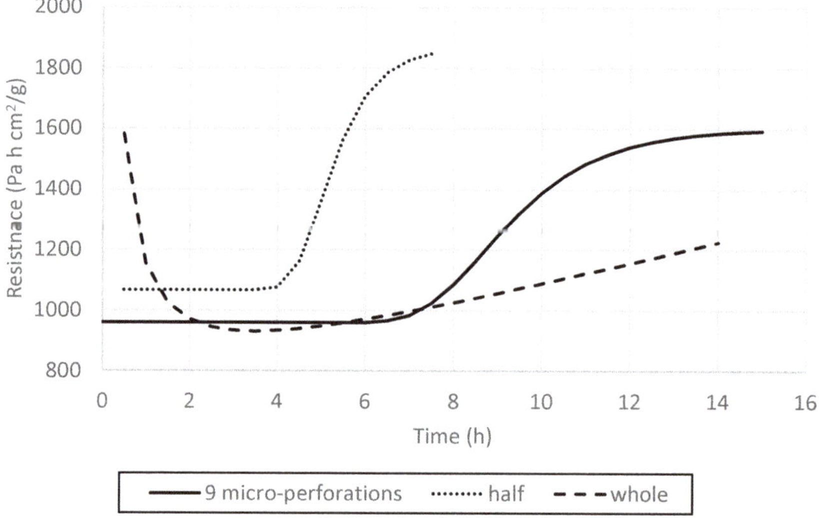

Figure 15. Comparison of mass-transfer resistance profiles as affected by the main treatments.

4. Conclusions

The outcomes from the present study demonstrate that CO_2 laser microperforation has high potential to be applied as a skin pretreatment for the freeze-drying of blueberries, which offers significant improvement on the process efficiency and final product quality.

The CO_2 laser treatment allowed the arrangement of microholes that minimally affect the quality and appearance of the fruit, serving as pathways for the flow of water vapor from the sublimating interface through the weakened mass transfer resistance of the berry skin, alleviating the pressure development under it, eventually avoiding the fruit bust, and enhancing the final product quality with a reduced processing time. This makes this novel technological approach an attractive alternative over traditional techniques.

Further research will be required to study other perforation arrangements and perforation depths in blueberries as well as other berry fruits during freeze-drying. The impact on energy consumption and cost-effectiveness should also be evaluated.

Author Contributions: S.A., P.M., J.U. and M.P. conceived, designed the study and wrote the manuscript; R.S., C.R., P.V., Contributed significantly to the design of the paper, edited and wrote some portion of the paper; All authors have read and agreed to the published version of the manuscript.

Funding: This research was funded by CONICYT through FONDECYT project number 1161675 (Sergio Almonacid), project number 1181270 (Ricardo Simpson).

Acknowledgments: We are grateful for the financial support provided by CONICYT through FONDECYT project number 1161675 (Sergio Almonacid), project number 1181270 (Ricardo Simpson), and project number.

References

1. St. George, S.D.; Cenkowski, S.; Muir, W.E. A review of drying technologies for the preservation of nutritional compounds in waxy skinned fruit. In Proceedings of the North Central ASAE/CSAE Conference Sponsored by the Manitoba Section of CSAE, Winnipeg, MB, Canada, 24–25 September 2004.

2. Stamenković, Z.; Pavkov, I.; Radojčin, M.; Horecki, A.T.; Kešelj, K.; Kovačević, D.B.; Putnik, P. Convective Drying of Fresh and Frozen Raspberries and Change of Their Physical and Nutritive Properties. *Foods* **2019**, *8*, 251. [CrossRef] [PubMed]

3. Mustafa, I.; Chin, N.L.; Fakurazi, S.; Palanisamy, A. Comparison of Phytochemicals, Antioxidant and Anti-Inflammatory Properties of Sun-, Oven- and Freeze-Dried Ginger Extracts. *Foods* **2019**, *8*, 456. [CrossRef] [PubMed]

4. Paredes-López, O.; Cervantes-Ceja, M.L.; Vigna-Pérez, M.; Hernández-Pérez, T. Berries: Improving Human Health and Healthy Aging, and Promoting Quality Life—A Review. *Plant Foods Hum. Nutr.* **2010**, *65*, 299–308. [CrossRef]

5. Han, J. Apparatus and Method for Perforations of Fruits and Vegetables. U.S. Patent 2011/0091620, 21 April 2011.

6. Ponting, J.D.; Mc Bean, D.M. Temperature and dipping treatment effects on drying rates and drying times of grapes, prunes and other waxy fruits. *Food Technol.* **1970**, *24*, 85–88.

7. Ponting, J.D. Osmotic dehydration of fruits, recent modifications and applications. *Process Biochem.* **1973**, *8*, 18–20.

8. Sunjka, P.S.; Raghavan, G.S.V. Assessment of pretreatment methods and osmotic dehydration for cranberries. *Can. Biosyst. Eng.* **2004**, *46*, 35–40.

9. Fujimaru, T.; Ling, Q.; Morrissey, M.T. Effects of Carbon Dioxide (CO2) Laser Perforation as Skin Pretreatment to Improve Sugar Infusion Process of Frozen Blueberries. *J. Food Sci.* **2012**, *77*, E45–E51. [CrossRef]

10. Azoubel, P.M.; Murr, F.E.X. Effect of pretreatment on the drying kinetics of cherry tomato (Lycopersicon esculentum var. cerasiforme). In *Transport Phenomena in Food Processing*; Welti-Chanes, J., Velez-Ruiz, F., Barbosa-Cánovas, G.V., Eds.; CRC Press: New York, NY, USA, 2003; pp. 137–151.

11. Feng HA, O.; Tang, J.; Mattinson, D.S.; Fellman, J.K. Microwave and spouted bed drying of frozen blueberries: The effect of drying and pre-treatment methods on physical properties and retention of flavour volatiles. *J. Food Process. Preserv.* **1999**, *23*, 463–479. [CrossRef]

12. Grabowski, S.; Marcotte, M.; Poirier, M.; Kudra, T. Drying characteristics of osmotically pretreated cranberries—Energy and quality aspects. *Dry. Technol.* **2002**, *20*, 1989–2004. [CrossRef]
13. Grabowski, S.; Marcotte, M. Pretreatment efficiency in osmotic dehydration of cranberries. In *Transport Phenomena in Food Processing*; Welti-Chanes, J., Velez-Ruiz, F., Barbosa-Cánovas, G.V., Eds.; CRC Press: New York, NY, USA, 2003; pp. 83–94.
14. Yang, A.P.; Wills, C.; Yang, T.C. Use of a Combination Process of Osmotic Dehydration and Freeze Drying to Produce a Raisin-Type Lowbush Blueberry Product. *J. Food Sci.* **1987**, *52*, 1651–1653. [CrossRef]
15. Ketata, M.; Desjardins, Y.; Ratti, C. Effect of liquid nitrogen pretreatments on osmotic dehydration of blueberries. *J. Food Eng.* **2013**, *116*, 202–212. [CrossRef]
16. Rudolph, K.S.; Edward, R. Process of Freeze-Drying Blueberries. U.S. Patent 3,467,530, 16 September 1969.
17. Yelden Eugene, F.; David, C. Food Processing Using Laser Radiation. U.S. Patent 124433A1, 29 May 2008.
18. Altshuler, G.; Mass, B.; Erofeev, A.V. Andrei. Laser Dermatology with Feedback Control. U.S. Patent 6O15404, 18 January 2000.
19. Bilanski, W.K.; Ferraz, A.C.O. Processing agricultural materials with a CO2 laser: A liner model. In Proceedings of the International Winter Meeting, Chicago, IL, USA, 17–20 December 1991; ASAE: Washington, DC, USA.
20. Ferraz, A.C.O.; Mittal, G.S.; Bilanski, W.K.; Abdullah, H.A. Mathematical modeling of laser based potato cutting and peeling. *BioSystems* **2007**, *90*, 602–613. [CrossRef]
21. Pikal, M.J.; Shah, S.; Senior, D.; Lang, J.E. Physical Chemistry of Freeze-drying: Measurement of Sublimation Rates for Frozen Aqueous Solutions by a Microbalance Technique. *J. Pharm. Sci.* **1983**, *72*, 635–650. [CrossRef] [PubMed]
22. Hua, T.; Liu, B.; Zhang, H. *Freeze—Drying of Pharmaceutical and Food Products*; CRC Press: Washington, DC, USA, 2010.
23. Milton, N.; Pikal, M.J.; Roy, M.L.; Nail, S.L. Evaluation of manometric temperature measurement as a method of monitoring product temperature during lyophilization. *PDA J. Pharm. Sci. Technol.* **1997**, *51*, 7–16.
24. Chouvenc, P.; Vessot, S.; Andrieu, J.; Vacus, P. Optimization of the Freeze-Drying Cycle: A New Model for Pressure Rise Analysis. *J. Dry. Technol.* **2004**, *22*, 1577–1601. [CrossRef]
25. Pikal, M.J. Use of Laboratory Data in Freeze Drying Process Design: Heat and Mass Transfer Coefficients and the Computer Simulation of Freeze Drying. *PDA J. Pharm. Sci. Technol.* **1985**, *39*, 115–139.
26. Kuu, W.Y.; Nail, S.L. Rapid freeze-drying cycle optimization using computer programs developed based on heat and mass transfer models and facilitated by tunable diode laser absorption spectroscopy (TDLAS). *J. Pharm. Sci.* **2009**, *98*, 3469–3482. [CrossRef]
27. Konstantinidis, A.K.; Kuu, W.; Otten, L.; Nail, S.L.; Sever, R.R. Controlled nucleation in freeze-drying: Effects on pore size in the dried product layer, mass transfer resistance, and primary drying rate. *J. Pharm. Sci.* **2011**, *100*, 3453–3470. [CrossRef]
28. Tang, X.; Nail, S.L.; Pikal, M.J. Evaluation of Manometric Temperature Measurement, a Process Analytical Technology Tool for Freeze-Drying: Part I, Product Temperature Measurement Submitted. *AAPS Pharm. SciTech* **2006**, *7*, 14. [CrossRef]
29. Matiacevich, S.; Cofré, D.C.; Silva, P.; Enrione, J.; Osório, F. Quality Parameters of Six Cultivars of Blueberry Using Computer Vision. *Int. J. Food Sci.* **2013**, *2013*, 419535. [CrossRef]
30. Lu, X.; Pikal, M.J. Freeze-Drying of Manitol-Trehalose-Sodium Chloride-Based Formulations: The Impact of Anealing on Dry Layer Resistance to Mass Transfer and Cake Structure. *Pharm. Dev. Technol.* **2004**, *9*, 85–95. [CrossRef] [PubMed]

Article

Freeze-Dried Gellan Gum Gels as Vitamin Delivery Systems: Modelling the Effect of pH on Drying Kinetics and Vitamin Release Mechanisms

Valentina Prosapio *, Ian T. Norton and Estefania Lopez-Quiroga

School of Chemical and Engineering, University of Birmingham, Birmingham B15 2TT, UK;
I.T.NORTON@bham.ac.uk (I.T.N.); E.Lopez-Quiroga@bham.ac.uk (E.L.-Q.)
* Correspondence: v.prosapio@bham.ac.uk

Received: 12 December 2019; Accepted: 5 March 2020; Published: 11 March 2020

Abstract: Freeze-dried gellan gum gels present great potential as delivery systems for biocompounds, such as vitamins, in food products. Here, we investigate the effect of modifying the gel pH—prior to the encapsulation process—on drying and release kinetics, and on delivery mechanisms from the substrate. Gellan gum gels were prepared at pH 5.2, 4 and 2.5 and loaded with riboflavin before being freeze-dried. Release tests were then carried out at ambient temperature in water. Five drying kinetics models were fitted to freeze-drying experimental curves using regression analysis. The goodness-of-fit was evaluated according to (i) the root mean squared error (ii), adjusted R-square (iii), Akaike information criterion (iv) and Bayesian information criterion. The Wang and Singh model provided the most accurate descriptions for drying at acidified pH (i.e., pH 4 and pH 2.5), while the Page model described better freeze-drying at pH 5.2 (gellan gum's natural pH). The effect of pH on the vitamin release mechanism was also determined using the Korsmeyer–Peppas model, with samples at pH 5.2 showing a typical Fickian behaviour, while acidified samples at pH 4 combined both Fickian and relaxation mechanisms. Overall, these results establish the basis for identifying the optimal conditions for biocompound delivery using freeze-dried gellan gels.

Keywords: freeze-drying; gellan gum; modified pH; riboflavin; drying kinetics; release mechanism; model discrimination

---

1. Introduction

Bioactive compounds used to enrich foods and beverages, such as vitamins, proteins or antioxidants, are highly sensitive to light, temperature and oxygen [1], undergoing degradation reactions (e.g., oxidation or pigment destruction) during processing that decrease their bioavailability [2]. To preserve them from degradation, those compounds can be encapsulated in suitable substrates according to the chosen functionalities [3] or required delivery rates; the release of a bioactive compound within the human body could be fast (mouth release) or prolonged over time (digestive tract release). The choice of the encapsulation technique is then key to preserving the biocompound and creating a suitable carrier microstructure—e.g., highly porous matrices can enhance mass transfer, leading to faster release rates. This makes freeze-drying a convenient technique for encapsulation of active biocompounds [4,5], as it helps with keeping the original porous structures of products, and its low temperature conditions also contribute to minimising degradation reactions [6,7].

One of the most versatile substrates employed in bioprocessing applications (i.e., food, pharma and healthcare technologies) is gellan gum gel. This is a non-toxic, biocompatible and biodegradable polymer [8] that has been extensively used as (i) a texturiser and gelling agent [9] in food applications; (ii) to formulate oral, nasal and ophthalmic formulations [10,11]; and (iii) as a scaffold for tissue regeneration [12,13]. A recent study [14] focused on the development of dried-gel structures from hydrocolloids has revealed the potential of gellan gum gels to be used as "controllable" carrier, showing that is possible to modulate the freeze dried-gel properties (i.e., target different microstructures and therefore different drying and rehydration kinetics) by modifying the pH of the initial gel solution.

To explore this promising path, this work focuses on the characterisation of freeze-dried gellan gum gels at different pHs as vitamin delivery systems. Freeze-drying kinetics, as well as release mechanisms and rates have been investigated using both empirical and modelling approaches. Gellan gum gels were prepared at different pHs (i.e., 5.2, 4 and 2.5) and then loaded with riboflavin (vitamin B2) before freeze-drying. Experimental drying curves were fitted to five common food drying models [15] (i.e., Newton, Page, Henderson and Pabis, logarithmic and Wang and Singh), and the effects of different pHs on freeze-drying kinetics were assessed. In addition, Information Theory criteria (Akaike and Bayesian information criteria) were used to discriminate the models [16] attending to their accuracy and complexity (i.e., number of parameters involved). The effect of pH on vitamin release (at ambient temperature) has been also studied using the classical Korsmeyer–Peppas model [17,18], and corresponding delivery mechanisms revealed. The findings of this work can help in the design of targeted freeze-dried gellan gum microstructures for the controlled delivery of active biocompounds in functional foods applications.

2. Materials and Methods

2.1. Materials

Low acyl (LA) gellan gum powder was provided by CPKelco (CPKelco, UK). Citric acid (purity 99%) and riboflavin (purity 98%) were supplied by Sigma-Aldrich (Sigma-Aldrich, Dorset, UK). All materials were used as received.

2.2. Preparation of Riboflavin-Loaded Gellan Gum Gels

Gellan gum powder was dissolved in distilled water at a concentration of 2% (*w/w*) and stirred at 85 °C for 2 h to ensure complete mixing [19]. This resulted in a gel solution at pH 5.2 (natural pH) (Seven compact Benchtop pH meter, Mettler Toledo UK). The gellan solutions at pH 4 and pH 2.5 were obtained by adding increasing quantities of citric acid at a 0.3 mol/L concentration [14]. All the solutions were placed in cylindrical moulds (diameter = 22 mm) for gelation [19]. Once the gels were formed, the moulds were cut into regular pieces (height = 15 mm,) and the resulting samples were soaked in a 2.7×10^{-4} mol/L riboflavin solution for 18 h. Finally, the loaded gels were washed with distilled water and blotted with paper to remove the vitamin settled on the surface.

2.3. Freeze Drying

The riboflavin-loaded gels were frozen at −20 °C for 24h and then freeze-dried at increasing processing times, from 2 h up to 18 h, in a bench top freeze dryer (SCANVAC Coolsafe™, model 110-4, Lynge, Denmark) with condenser temperature of −110 °C and chamber pressure of 10 Pa [14]. The experiments were performed in triplicates, and each batch of freeze-dried samples was weighted to measure the water content.

2.4. Water Activity Analysis

Water activity (a_w) of wet and dried gels was measured using an AquaLab® dew point water activity meter (model 4 TE, Decagon Devices Inc., Pullman, WA, USA). The temperature-controlled sample chamber was set to 25 °C [14,15]. All analyses were carried out in triplicate.

2.5. Vitamin Release Experiments

Vitamin delivery analyses were performed using a UV–Vis spectrophotometer (Orion Aquamate, Thermo Scientific, UK) at 444 nm. The loaded gels were placed in stirred (250 rpm) distilled water (200 mL) at room temperature. To measure vitamin release, aliquots of 3 mL were withdrawn from the release medium, analysed with the spectrophotometer and poured back into the release medium. The vitamin content in the release medium was then expressed as normalised vitamin release (NVR), a dimensionless quantity defined as:

$$NVR = \frac{V(t)}{V_{total}} \tag{1}$$

All analyses were carried out in triplicate.

2.6. Drying Kinetics

The kinetics of moisture loss during freeze-drying of the loaded gellan gum gels were described by five empirical models commonly employed to characterise drying kinetics in foods [15,16]: Newton, Page, Henderson and Pabis, logarithmic and Wang and Singh. Table 1 lists them all alongside their expressions.

Table 1. Drying kinetics models considered in this work to describe moisture loss during freeze-drying of riboflavin-loaded gellan gum gels with different pHs.

Drying model	Expression [15,16]	
Newton	$MR = e^{-k_1 t}$	Table Equation (1)
Page	$MR = e^{-k_2 t^n}$	Table Equation (2)
Henderson and Pabis	$MR = a_1 e^{-k_3 t}$	Table Equation (3)
Logarithmic	$MR = a_2 e^{-k_4 t} + b_1$	Table Equation (4)
Wang and Singh	$MR = 1 + k_5 t + k_6 t^2$	Table Equation (5)

Parameter units: (h^{-1}) for k_1, k_3, k_4, k_5; (h^{-n}) for k_2; (h^{-2}) for k_6; a_1, a_2 and b_1 are dimensionless.

To fit the models to the experimental freeze-drying curves, the (dimensionless) moisture ratios (MRs) of the samples were calculated first from the measured water content data as follows [15]:

$$MR = \frac{X(t) - X_{eq}}{X_0 - X_{eq}} \tag{2}$$

where $X(t)$ is the moisture content on a dry basis for the different processing times (h), X_0 is the initial moisture content (*w/w* d.b.) and X_{eq} is the equilibrium moisture content (*w/w* d.b). The equilibrium moisture content for the dehydrated gels was calculated using the GAB model with measured water activities and parameters for gellan gums presented in [9]:

$$\frac{a_w}{X_{eq}} = 0.165 + 14.3a_w - 13.2\,a_w{}^2 \tag{3}$$

For all the models in Table 2, the unknow parameters (parameters a_j and k_i, with $j = 1,2$ and $i = 1, \dots, (6)$ were estimated using regression analysis. The error (e) between the experimental (θ) and predicted (i.e., fitted) MR values ($\bar{\theta}$) [15],

$$J = \sum_i^N e_i{}^2 = \sum_i^N \left(\theta_i - \bar{\theta}_i\right)^2, \tag{4}$$

was minimised for all the i measurements that formed the experimental data set of size N using a nonlinear least squares method (implemented in Matlab with tolerance 10^{-10}).

Table 2. Regression and goodness-of-fit results for the drying kinetics models.

Model		Parameters	RMSE	R^2_{adj}	BIC	AICC
Newton						
	pH 2.5	$k_1 = 0.225$	0.066	0.973	−13.71	−14.92
	pH 4	$k_1 = 0.157$	0.038	0.988	−20.18	−21.39
	pH 5.2	$k_1 = 0.176$	0.076	0.961	−11.94	−13.15
Page						
	pH 2.5	$k_2 = 0.117; n = 1.417$	0.035	0.990	−20.61	−16.20
	pH 4	$k_2 = 0.132; n = 1.101$	0.035	0.988	−20.76	−16.34
	pH 5.2	$k_2 = 0.071; n = 1.551$	0.04	0.986	−19.10	−14.69
Henderson						
	pH 2.5	$a_1 = 1.036; k_3 = 0.232$	0.064	0.968	−13.57	−9.15
	pH 4	$a_1 = 1.006; k_3 = 0.158$	0.038	0.985	−19.63	−15.21
	pH 5.2	$a_1 = 1.043; k_3 = 0.185$	0.073	0.955	−11.87	−7.45
Logarithmic						
	pH 2.5	$a_2 = 1.146; k_4 = 0.186; b_2 = -0.121$	0.042	0.981	−16.69	−4.07
	pH 4	$a_2 = 1.114; k_4 = 0.126; b_2 = -0.125$	0.021	0.994	−24.95	−12.32
	pH 5.2	$a_2 = 1.142; k_4 = 0.153; b_2 = -0.109$	0.064	0.954	0.89	−11.73
Wang and Singh						
	pH 2.5	$k_7 = -0.160; k_8 = 0.0056$	3.52×10^{-4}	0.999	−37.85	−33.44
	pH 4	$k_7 = -0.122; k_8 = 0.0037$	0.005	0.990	−17.37	−21.78
	pH 5.2	$k_7 = -0.138; k_8 = 0.0045$	0.045	0.983	−13.33	−17.75

Parameter units: (h^{-1}) for k_1, k_3, k_4, k_5; (h^{-n}) for k_2; (h^{-2}) for k_6; a_1, a_2 and b_1 are dimensionless.

The goodness-of-fit of each fitted model was then assessed using three statistical measures that take into account the complexity (i.e., number of parameters, p) of each model [20]. These were:

-   The adjusted R^2 [15]:

$$R^2_{adj} = 1 - \frac{N-1}{N-p}\left(1 - R^2\right)$$
(5)

where R^2 is the regression coefficient of determination.
-   The corrected Akaike information criterion (*AICC*) [15,21]:

$$AICC = AIC + \frac{2p(p+1)}{N-p-1}$$
(6)

where *AIC* is the Akaike information criterion [21,22].
-   The Bayesian information criterion (*BIC*) [20]:

$$BIC = p\ln(N) - 2\,\ln(L)$$
(7)

where L is the maximum log-likelihood of the estimated model.

The goodness of the fit (or the likelihood) can be increased by adding more parameters to the model. However, this will increase complexity and might result in overfitting (i.e., more parameters than can be estimated with the available data), all which is penalized with higher AICC and BIC values [15,20]. Therefore, the model with best performance will be the one with higher R^2_{adj} and lower *AICC* and *BIC* values [20].

2.7. Kinetics and Mechanisms of Vitamin Release

The Korsmeyer–Peppas model [17] has been used to describe release kinetics and identify delivery mechanisms. It is a semi-empirical power law that relates the fractional release of vitamin/drug to the release time [17,18]:

$$\frac{M(t)}{M_\infty} = k_{rel}t^{n_{rel}}$$
(8)

where the $M(t)$ and M_∞ are the cumulative amounts of drug released at time t (measured in hours h) and infinite time, respectively; the constant k_{rel} (in $h^{-n_{rel}}$ units) relates to the structure and geometry of the delivery matrix (in this case, the freeze-dried gels); and the dimensionless exponent n_{rel} is the release mechanism indicator. For cylindrical substrates $n_{rel} \leq 0.45$ defines Fickian mechanisms, while anomalous/non-Fickian delivery is described by $n_{rel} > 0.45$ [17,18]. Experimental release curves were fitted to Equation (8) using a nonlinear least squares method [15], and parameters k_{rel} and n_{rel} were estimated within 95% confidence intervals.

3. Results and Discussion

3.1. Effect of pH on Moisture Losses during Freeze-Drying

Figure 1 shows a comparison of the drying curves, in terms of the moisture ratio (MR) and freeze-drying processing times (h), for gellan gum gels at pH 2.5, 4 and 5.2 (natural) loaded with riboflavin. Gels at pH 2.5 exhibit the fastest drying rates, with most of the moisture content (MR~ 0.25) removed during the first 6 h of the freeze-drying process. On the other hand, samples at pH 5.2 and 4 followed a very similar drying trend up to the first 4 h of processing. From this time onwards, the samples at pH 4 present a significantly slower drying rate; i.e., MR~ 0.3 at t = 8 h compared to MR~ 0.15 for pH 5.2 at the same time. The three samples were completely dried (i.e., free water totally removed) by the end of the freeze-drying experiments at t= 18 h, independent of their pHs.

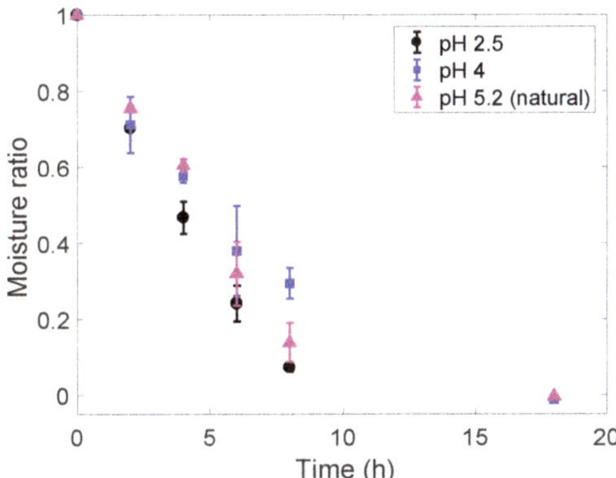

Figure 1. Moisture ratio evolution along time for 2% (*w/w*) gellam gums with pH 2.5 (black dots), pH 4 (blue squares) and pH 5.2 (magenta triangles) loaded with riboflavin during the conducted freeze-drying experiments.

The fastest drying rates observed for the gels at lower pHs can be explained by the effects of acidifying the gel solution. As reported by Cassanelli et al., [14], the acidification step both enhances ice crystal nucleation and weakens the gel structure at pH values as low as pH 2.5. The combination of these two effects might favour a more interconnected pore structure—i.e., more nuclei will lead to more crystals that will find lower resistance in the weak gel structure to form a network. This could lead to faster drying rates and also affect the strength of the rehydrated structure.

3.2. Freeze-Drying Kinetics: Parameter Estimation and Model Discrimination

Estimated parameters for all the drying models considered in this work (i.e., Newton, Page, Henderson and Pabis, logarithmic and Wang and Singh) are listed in Table 2, together with the RMSE (root mean square error) for each fitting and the results corresponding to the goodness-of-fit of each model. According to these results, the models that provide more accurate descriptions for the drying kinetics are the Page (Table Equation 1) and the Wang and Singh (Table Equation 5) models, both presenting $R^2_{adj} \sim 0.99$ (in average) for all pH values.

For samples at modified pHs (i.e., pH 2.5 and 4), the Wang and Singh model presents the lowest RMSE (3.52×10^{-4} for pH 2.5) and the highest R^2_{adj}, while for the freeze-dried gels at pH 5.2, the Page model is the model that presents the best fitting (RMSE = 0.040 and $R^2_{adj} = 0.986$). This is in agreement with fittings reported in Cassanelli et al. [19], which showed the Page model as the best option to describe the freeze-drying kinetics of non-loaded gellan gels at natural pH.

The goodness-of-fit for all models is illustrated in Figure 2, where experimental values are plotted against predicted moisture ratios for each drying model at all pH studied. This graph also shows the accuracy of the Page and Wang models, for which most of the predicted points lie on the correlation line (see Figure 2b,d).

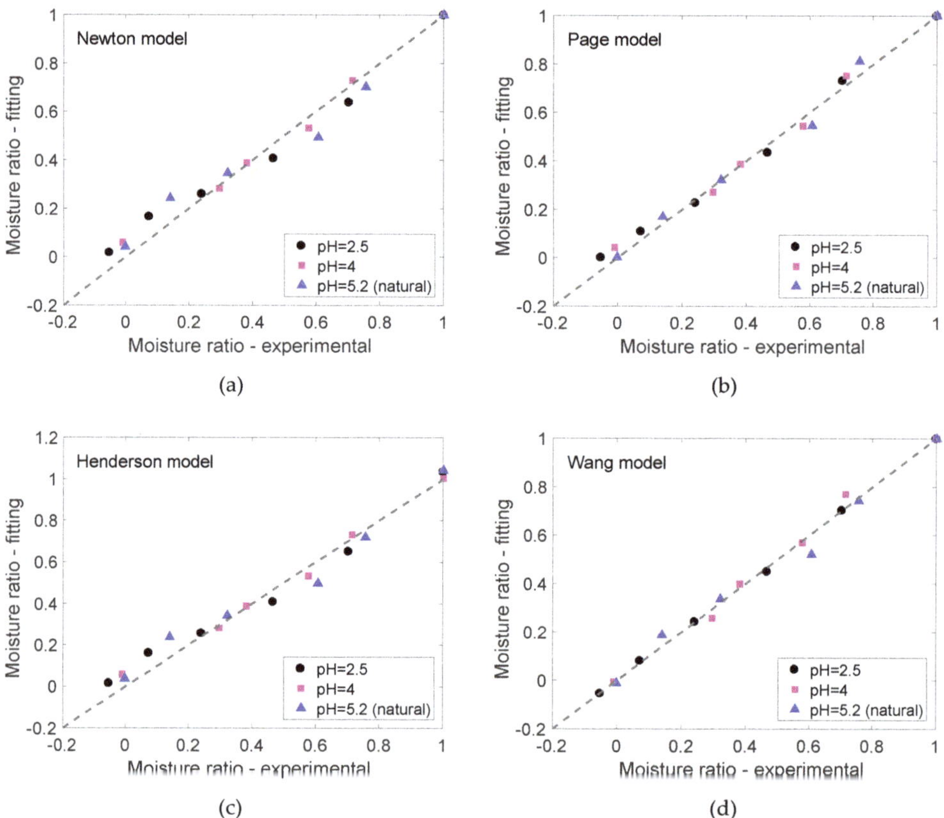

Figure 2. Correlation between predicted and experimental moisture contents for freeze-dried 2% (w/w) gellan gum samples for: (**a**) Newton model (Table Equation 1), (**b**) Page model (Table Equation 2), (**c**) Henderson and Pabis model (Table Equation 3) and (**d**) Wang model (Table Equation 5).

When comparing models with similar accuracies, the *AICC* criterion constitutes the best measure to discriminate models, with more negative *AICC* values indicating better model performances. According to this, if the Newton and Page models were compared at pH 4—the pH at which both models show very similar RMSE, R^2_{adj} and *BIC*—the lower AICC (−21.39) of the Newton model would make it the preferred one. This criterion is also an indicator of the complexity (e.g., number of parameters) of the assessed models—the Newton model involves a single parameter (k_1), compared to the two needed in the Page model (k_2, n). On the other hand, the logarithmic model (Table Equation 4), with the highest number of parameters considered ($p = 3$), presents the less negative *AICC* values at each pH.

Effect of pH on the Drying Kinetic Parameters

The effect of pH on drying kinetic parameters has been determined by analysing the values of the constants in Newton (Table Equation 1) and Henderson and Pabis (Table Equation 3) models. These two models are derived from Newton's cooling law and Fick's Second law [16], respectively, so their constants enclose physical meaning—as opposite to the Page and Wang and Singh models that are purely empirical [16].

Parameters k_1 (Newton) and k_3 (Henderson) in Table 2, both time constants (h^{-1}), characterise the drying rates of the system, while a_1 (Henderson) is a dimensionless parameter related to the shape and structural properties of the samples [16].

For gellan gum gels at different pHs, both Newton and Henderson rate parameters (i.e., k_1 and k_3, respectively) show very similar trends. The higher values ($k_1 = 0.225\,h^{-1}$ and $k_3 = 0.232\,h^{-1}$) correspond to samples at pH 2.5, indicating a faster dehydration process. On the other hand, rate constants for samples at pH 4 are the lowest ($k_1 = 0.157\,h^{-1}$ and $k_3 = 0.158\,h^{-1}$), which relates to the slower drying rate of these samples. This is in agreement with differences on moisture ratios (MR) at different pHs discussed in Section 3.1.

The values of constant a_1 are again similar for samples at pH 2.5 and 5.2 ($a_1 = 1.036$ and $a_1 = 1.043$), which suggests no significant structural differences at those pH values. However, the value of a_1 for freeze-dried gels at pH 4 ($a_1 = 1.006$) suggests changes in microstructure that might be behind the different drying rates observed at this particular pH. These findings are in agreement with the mechanical properties (i.e., higher gel strength and Young's modulus) reported in Cassanelli et al. [14] for gellan gels at pH 4 before and after freeze-drying—"stronger" gels might make ice nucleation and growth difficult, and therefore affect the freeze-dried microstructures of the gels.

3.3. Riboflavin Release from Freeze-Dried Gellan Gels at Different pHs

Figure 3 presents experimental riboflavin release curves from freeze-dried gels prepared at different pHs, plotted as normalised vitamin released (NVR) across time. Data in this graph show significant differences in release times: freeze-dried gels at pH 4 completed the vitamin release in approximately 9.5 h; gels at natural pH (5.2) were fully unloaded after 6h, and total vitamin delivery took 3h for gels at pH 2.5. Samples at pH 2.5 presented a weak structure—in accordance with strength at fresh and freezing stages—that lead to breakage during the release experiments. This increased the surface area of the gels in contact with the release medium, which explains the shorter delivery times.

The observed differences in the riboflavin release times to the medium can be related to the different microstructures and mechanical properties of the gels. Both Norton et al. [23] and Cassanelli et al. [14] reported that freeze-dried gellan gum gels at pH 4 exhibited an aggregated and rigid structure. This can impede mass transfer within the gel, increasing the time needed to release the vitamin completely from the substrate and leading to longer delivery processes. A much lower level of aggregation and very weak structures were reported for freeze-dried gels at pH 2.5 [14,23], which is also in agreement with our experimental observations. According to the same authors, unloaded freeze-dried gels at natural pH (pH 5.2) exhibit intermediate levels of aggregation [14,24], explaining the also intermediate release times shown in Figure 3.

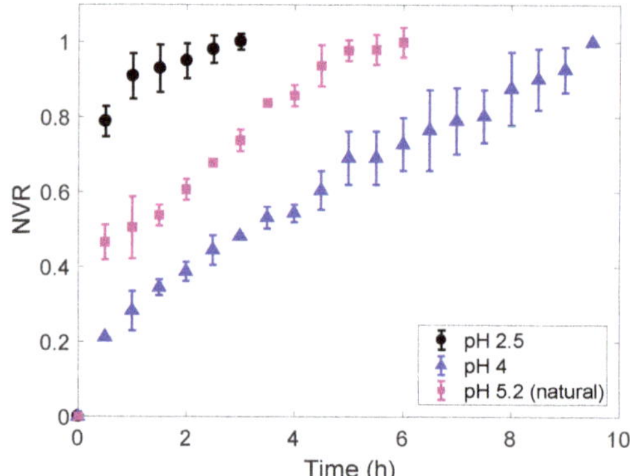

Figure 3. Release curves for the riboflavin encapsulated in freeze-dried 2% (*w/w*) gellan gums with different pHs. The vitamin content in the release medium is expressed as normalised vitamin released (NVR). Error bars correspond to triplicate tests.

3.4. Delivery Mechanisms at Different pHs

To estimate the value of the dimensionless parameter n_{rel} in Equation (8), and therefore determine the release mechanism governing riboflavin delivery, the portion of the release curves (Figure 3) corresponding to the first 60% of the total released vitamin—i.e., release curve portions such that $\frac{M(t)}{M_\infty} = NRV \leq 0.60$—were fitted to the Korsmeyer–Peppas model [17,18]. Samples at pH 2.5 were not considered in this analysis, as they broke into several pieces during the release tests, leading to delivery conditions out of the scope of this work. Table 3 lists the parameters k_{rel} and n_{rel} (95% CI) estimated at pH 5.2 and pH 4 (see Table 3 for parameter units). These results are discussed next.

Table 3. Fitted parameters (95% CI) for the Korsmeyer–Peppas release model and release mechanisms found.

	k_{rel}	n_{rel}	Release Mechanism
pH 2.5	-	-	-
pH 4	0.287 (0.277, 0.297)	0.472 (0.441, 0.504)	Anomalous transport
pH 5.2 (natural)	0.509 (0.502, 0.515)	0.131 (0.102, 0.161)	Fickian diffusion

Parameter units: $(h^{-n_{rel}})$; n_{rel} is dimensionless.

3.4.1. Release from Freeze-Dried Gellan Gels at pH 5.2

Riboflavin delivery from gels at natural pH (pH 5.2) is characterised by a shape constant $k_{rel} = 0.509$ diffusional coefficient $n_{rel} = 0.131$ (see Table 3 for the corresponding confidence intervals). According to the classification given in [21,22], this indicates that the governing release mechanism is purely Fickian, as $n_{rel} = 0.131 < 0.45$, which is the limiting value of the diffusional coefficient for Fickian transport mechanisms. Therefore, we can define an apparent diffusion coefficient D_{app} (m^2/s) for samples at pH 5.2 using a short-time approximation of Fick's Second law [18]:

$$\frac{M(t)}{M_\infty} = 4\left[\frac{D_{app}t}{\pi a^2}\right]^{\frac{1}{2}} \tag{9}$$

As the aspect ratio of the cylindrical gels is approximately in the order of 1, the predictive capabilities of the short-time solution include up to the 85% of the total vitamin release [22]. Thus, values such that $\frac{M(t)}{M_\infty} = NRV \leq 0.85$ in the release curve at pH 5.2 were fitted to Equation (9). This gave an estimated D_{app} = 1.325 × 10^{-9} m^2/s with 95% CI defined by [1.086 × 10^{-9}, 1.564 × 10^{-9}] m^2/s. Figure 4 presents the comparison between the experimental and predicted release curve using D_{app}, showing a good agreement between the modelled Fickian delivery mechanism and the experimental data—and thus confirming Fickian transport for riboflavin released from freeze-dried gellan gels at pH 5.2

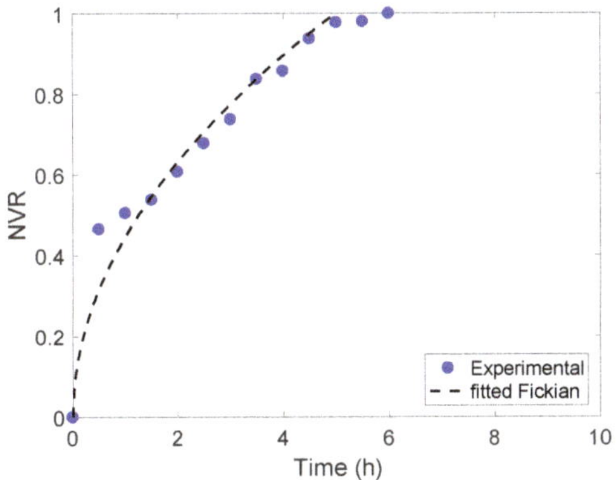

Figure 4. Predicted release curve for encapsulated riboflavin at pH 5.2 using estimated D_{app} (dash –) compared to experimental curve (blue dots).

3.4.2. Release from Freeze-Dried Gellan Gels at pH 4

As shown in Table 3, the estimated shape constant and diffusional exponent at pH 4 were k_{rel} = 0.287 and n_{rel} = 0.472, respectively. These estimates (i) confirm the structural difference of samples at pH 4 discussed in previous subsections, as the value of k_{rel} at pH 4 is almost half of the corresponding to pH 5.2, and (ii) indicate an anomalous delivery mechanism from freeze-dried gellan matrices at pH 4, since we estimated n_{rel} > 0.45.

Anomalous mass transport can be defined as a mix between Fickian and non-Fickian mechanisms, for which the general form of the Korsmeyer–Peppas model presented in Equation (8) can be split into two contributions [24]:

$$\frac{M(t)}{M_\infty} = k_{rel}t^{n_{rel}} = k_1^{rel}t^{m_{rel}} + k_2^{rel}t^{2m_{rel}}. \tag{10}$$

The first one ($k_1^{rel}t^m$) represents the Fickian part, while the second term ($k_2^{rel}t^{2m}$) accounts for the relaxational contribution [24], which is related to stresses and state transitions of polymeric matrices. The dimensionless diffusional exponent m_{rel} in Equation (10) can be determined from aspect ratio (i.e., height/diameter of the sample) correlations. For the ratio characterising the cylindrical samples used here (~1.5), m_{rel} = 0.43 [24].

Estimated values for both k_1^{rel} = 0.264 (0.240, 0.289) and k_2^{rel} = 0.022 (0.007, 0.038) were then obtained by fitting Equation (10) to the experimental release curves at pH 4 for NMC <0.60 with 95% confidence intervals (values in the parenthesis). Units for k_1^{rel} and k_2^{rel} are ($h^{-m_{rel}}$). These estimates give an idea of the relevance of each contribution. For samples at pH 4, $k_1^{rel} \gg k_2^{rel}$, showing that the release of riboflavin from the freeze-dried gels at pH 4 is mostly Fickian. This is confirmed by the vitamin release percentages due to each contribution calculated as Peppas and Sahlin [24]:

$$P_{Fickian} = \frac{1}{1 + \frac{k_2^{rel}}{k_1^{rel}}t^{m_{rel}}} \; ; \; P_{non-Fick} = 100 - P_{Fickian} = P_{Fickian}\frac{k_2^{rel}}{k_1^{rel}}t^{m_{rel}} \tag{11}$$

They are presented in Figure 5a. Overall, the Fickian contribution is predominant along the release process, i.e., overall $P_{Fickian} > 80\%$, with values closer to 90% at the initial times of the delivery test, while relaxation effects are more important towards the end of the experiment, as the delivery of riboflavin is closer to completion.

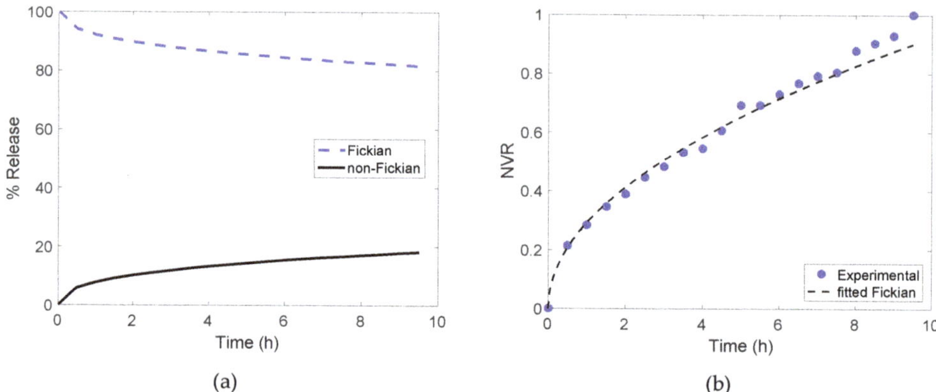

Figure 5. (**a**) Fickian and non-Fickian release percentages for riboflavin corresponding to sample with pH 4 when anomalous transport mechanism was considered. (**b**) Predicted release curve for encapsulated riboflavin at pH 4 considering pure Fickian mechanism and estimated D_{app} (dash –) compared to experimental curve (blue dots).

Given the relevance of Fickian transport during the release process at pH 4, and with the estimated diffusional exponent n_{rel} so close to the Fickian limiting value of 0.45—confidence intervals for this parameter are actually cross-boarding this limit, i.e., (0.441, 0.504) as shown in Table 3—a hypothetical pure Fickian riboflavin delivery at pH 4 has been also assessed. Following the procedure explained in Section 3.4.1, an apparent diffusion coefficient $D_{app} = 5.626 \times 10^{-10}$ m^2/s was estimated within a 95% confidence interval $(5.409 \times 10^{-10}, 5.842 \times 10^{-10})$ m^2/s. This estimate together with the short time approximately described in Equation (9) was used to obtain a predicted release curve, which is presented in Figure 5b alongside the experimentally obtained curve. As the comparison reveals, the hypothetical pure Fickian mechanism describes the behaviour observed during the release tests well, and it could be used to predict riboflavin delivery—neglecting relaxation effects—from freeze-dried gellan gels at pH 4 with high accuracy.

4. Conclusions

This work demonstrates the potential to control biocompound release from freeze-dried gellan gum gels by modifying the pH of the substrate during gel formation, and prior to the encapsulation stage. As an exemplar of a relevant biocompound, riboflavin (i.e., vitamin B$_2$) was used.

Freeze-drying kinetics, as well as release mechanisms, were experimentally investigated and modelled. Five different drying kinetics models were discriminated by accuracy and goodness-of-fit using statistical measures (i.e., RMSE, R^2_{adj}, AICC and BIC). For samples at natural pH (pH 5.2), the Page model provided the most accurate description of freeze-drying kinetics, while the Wang and Singh model predicted more accurately, the kinetics at acidified pH (i.e., 4 and 2.5).

Results revealed consistent differences in the behaviour of substrates at pH 4. Such differences reflect slower drying and release kinetics, as well as a different delivery mechanism—samples at natural pH (pH 5.2) exhibited Fickian transport, while acidified samples (pH 4) were characterised by an anomalous release mechanism, but with a predominantly Fickian contribution (80–90%).

Overall, this work shows the potential of modified pH freeze-dried gellan gum gel matrices for controlled riboflavin release, demonstrating that:

(i)   These hydrogels could be used in different enriched food and/or beverage products.
(ii)   Model-based approaches like the one presented here represent useful tools for the design of novel food formulations.

Author Contributions: Conceptualization, V.P., E.L.-Q.; Experimental work and writing, V.P.; Modelling work and writing, E.L.-Q. Supervision and resources, I.T.N. All authors have read and agreed to the published version of the manuscript.

Funding: This research was funded by EPSRC grant no. EP/S023070/1.

Acknowledgments: Authors acknowledge financial support from the EPSRC Centre for Doctoral Training in Formulation Engineering (grant number EP/S023070/1) as well as support from the Centre's Director, Prof. Fryer.

Conflicts of Interest: The authors declare no conflict of interest

References

1. Sheraz, M.A.; Kazi, S.H.; Ahmed, S.; Anwar, Z.; Ahmad, I. Photo, thermal and chemical degradation of riboflavin. *Beilstein J. Org. Chem.* **2014**, *10*, 1999–2012. [CrossRef] [PubMed]
2. Moreno, J.J. *Innovative Processing Technologies for Foods with Bioactive Compounds*; CRC Press: Boca Raton, FL, USA, 2016.
3. De Marco, I.; Reverchon, E. Starch aerogel loaded with poorly water-soluble vitamins through supercritical CO2 adsorption. *Chem. Eng. Res. Des.* **2017**, *119*, 221–230. [CrossRef]
4. Fang, Z.; Bhandari, B. Spray drying, freeze drying and related processes for food ingredient and nutraceutical encapsulation. In *Encapsulation Technologies and Delivery Systems for Food Ingredients and Nutraceuticals*; Woodhead Publishing: Cambridge, UK, 2012.
5. Oikonomopoulou, V.P.; Krokida, M.K. Freeze drying and microwave freeze drying as encapsulation methods. *Therm. Nontherm. Encapsul. Methods* **2017**, 39–65. [CrossRef]
6. Karam, M.C.; Petit, J.; Zimmer, D.; Baudelaire Djantou, E.; Scher, J. Effects of drying and grinding in production of fruit and vegetable powders: A review. *J. Food Eng.* **2016**, *188*, 32–49. [CrossRef]
7. Reyes, A.; Mahn, A.; Cares, V. Analysis of dried onions in a hybrid solar dryer, freeze dryer and tunnel dryer. *Chem. Eng. Trans.* **2015**, *43*, 139–144.
8. Mahajan, H.S.; Gattani, S.G. Gellan Gum Based Microparticles of Metoclopromide Hydrochloride for Intranasal Delivery: Development and Evaluation. *Chem. Pharm. Bull.* **2009**, *57*, 388–392. [CrossRef]
9. Abramovič, H.; Klofutar, C. Water adsorption isotherms of some gellan gum samples. *J. Food Eng.* **2006**, *77*, 514–520. [CrossRef]
10. Agnihotri, S.A.; Jawalkar, S.S.; Aminabhavi, T.M. Controlled release of cephalexin through gellan gum beads: Effect of formulation parameters on entrapment efficiency, size, and drug release. *Eur. J. Pharm. Biopharm.* **2006**, *63*, 249–261. [CrossRef]
11. Osmałek, T.; Froelich, A.; Tasarek, S. Application of gellan gum in pharmacy and medicine. *Int. J. Pharm.* **2014**, *466*, 328–340. [CrossRef]
12. Grenier, J.; Duval, H.; Barou, F.; Lv, P.; David, B.; Letourneur, D. Mechanisms of pore formation in hydrogel scaffolds textured by freeze-drying. *Acta Biomater.* **2019**, *94*, 195–203. [CrossRef]
13. Kim, D.; Thangavelu, M.; Cheolui, S.; Kim, H.S.; Choi, M.J.; Song, J.E.; Khang, G. Effect of different concentration of demineralized bone powder with gellan gum porous scaffold for the application of bone tissue regeneration. *Int. J. Biol. Macromol.* **2019**, *134*, 749–758. [CrossRef] [PubMed]
14. Cassanelli, M.; Prosapio, V.; Norton, I.; Mills, T. Acidified/basified gellan gum gels: The role of the structure in drying/rehydration mechanisms. *Food Hydrocoll.* **2018**, *82*, 346–354. [CrossRef]
15. Onwude, D.I.; Hashim, N.; Janius, R.B.; Nawi, N.M.; Abdan, K. Modeling the Thin-Layer Drying of Fruits and Vegetables: A Review. *Compr. Rev. Food Sci. Food Saf.* **2016**, *15*, 599–618. [CrossRef]
16. Lopez-Quiroga, E.; Prosapio, V.; Fryer, P.J.; Norton, I.T.; Bakalis, S. Model discrimination for drying and rehydration kinetics of freeze-dried tomatoes. *J. Food Process Eng.* **2019**, e13192. [CrossRef]

17. Peppas, N.A.; Narasimhan, B. Mathematical models in drug delivery: How modeling has shaped the way we design new drug delivery systems. *J. Control. Release* **2014**, *190*, 75–81. [CrossRef]
18. Ritger, P.L.; Peppas, N.A. A simple equation for description of solute release I. Fickian and non-fickian release from non-swellable devices in the form of slabs, spheres, cylinders or discs. *J. Control. Release* **1987**, *5*, 23–36. [CrossRef]
19. Cassanelli, M.; Norton, I.; Mills, T. Role of gellan gum microstructure in freeze drying and rehydration mechanisms. *Food Hydrocoll.* **2018**, *75*, 51–61. [CrossRef]
20. Spiess, A.-N.; Neumeyer, N. An evaluation of R 2 as an inadequate measure for nonlinear models in pharmacological and biochemical research: A Monte Carlo approach. *BMC Pharmacol.* **2010**, *10*, 6. [CrossRef]
21. Moxon, T.E.; Nimmegeers, P.; Telen, D.; Fryer, P.J.; Van Impe, J.; Bakalis, S. Effect of chyme viscosity and nutrient feedback mechanism on gastric emptying. *Chem. Eng. Sci.* **2017**, *171*, 318–330. [CrossRef]
22. Akaike, H. A new look at the statistical model identification. *IEEE Trans. Autom. Control* **1974**, *19*, 716–723. [CrossRef]
23. Norton, A.B.; Cox, P.W.; Spyropoulos, F. Acid gelation of low acyl gellan gum relevant to self-structuring in the human stomach. *Food Hydrocoll.* **2011**, *25*, 1105–1111. [CrossRef]
24. Peppas, N.A.; Sahlin, J.J. A simple equation for the description of solute release. III. Coupling of diffusion and relaxation. *Int. J. Pharm.* **1989**, *57*, 169–172. [CrossRef]

Article

Chickpea Cultivar Selection to Produce Aquafaba with Superior Emulsion Properties

Yue He [1], Youn Young Shim [2,3,4,5], Rana Mustafa [2,3,4], Venkatesh Meda [1] and Martin J.T. Reaney [2,3,4,*]

1   Department of Chemical and Biological Engineering, University of Saskatchewan, Saskatoon, SK S7N 5A9, Canada; yuh885@mail.usask.ca (Y.H.); venkatesh.meda@usask.ca (V.M.)
2   Department of Plant Sciences, University of Saskatchewan, Saskatoon, SK S7N 5A8, Canada; younyoung.shim@usask.ca (Y.Y.S.); rana.mustafa@usask.ca (R.M.)
3   Prairie Tide Diversified Inc., Saskatoon, SK S7J 0R1, Canada
4   Guangdong Saskatchewan Oilseed Joint Laboratory, Department of Food Science and Engineering, Jinan University, Guangzhou, Guangdong 510632, China
5   Department of Integrative Biotechnology, Sungkyunkwan University, Suwon, Gyeonggi-do 16419, Korea
*   Correspondence: martin.reaney@usask.ca; Tel.: +1-306-9665027

Received: 16 November 2019; Accepted: 13 December 2019; Published: 15 December 2019

Abstract: Aquafaba (AQ), a viscous by-product solution produced during cooking chickpea or other legumes in water, is increasingly being used as an egg replacement due to its ability to form foams and emulsions. The objectives of our work were to select a chickpea cultivar that produces AQ with superior emulsion properties, and to investigate the impact of chickpea seed physicochemical properties and hydration kinetics on the properties of AQ-based emulsions. AQ from a Kabuli type chickpea cultivar (CDC Leader) had the greatest emulsion capacity (1.10 ± 0.04 m^2/g) and stability ($71.9 \pm 0.8\%$). There were no correlations observed between AQ emulsion properties and chickpea seed proximate compositions. Meanwhile, AQ emulsion properties were negatively correlated with AQ yield and moisture content, indicating that AQ with higher dry-matter content displayed better emulsion properties. In conclusion, the emulsification properties of aquafaba are greatly influenced by the chickpea genotype, and AQ from the CDC Leader chickpea produced the most stable food oil emulsions.

Keywords: aquafaba; chickpea; emulsifiers; egg replacer; egg-free products

---

1. Introduction

Food oil emulsions are significant components of food. Whole egg, egg yolk and egg white are typical ingredients in a range of food oil emulsions, such as mayonnaise and salad dressing, as these materials are efficient, natural emulsifiers [1] for a variety of oil/water (O/W) and water/oil (W/O) emulsions. The high emulsifying capacity of egg is related to the phospholipids (lecithin), lipoproteins (low-density lipoproteins, and high-density lipoproteins) and non-associated proteins (livetin and phosvitin) [2,3]. These proteins have amphiphilic properties and act as surface-active substances in multiphase systems, such as mayonnaise.

Unfortunately, egg products are one of the more frequent agents associated with food allergies, especially in infants and young children [4]. Egg allergens are mainly present in egg white. Ovalbumin, constituting 54% of egg white protein [4], is one of the major egg allergens [5]. In addition, egg is not suitable for consumers with special dietary restrictions, and those that cannot eat egg for religious reasons or personal lifestyle choices [6]. Moreover, egg yolk contains cholesterol (5.2% of total lipid) which is linked to cardiovascular disease.

Although a cholesterol limit is not mentioned in the 2015–2020 Dietary Guideline for Americans, it is still recommended that the elderly and people with previous incidents of heart disease limit their dietary cholesterol intake [6]. Furthermore, a segment of consumers cite environmental concerns related to egg production as a rationale to avoid egg consumption [7]. Therefore, many scientists and food processing companies are developing innovative new egg replacements to cater to a growing demand for egg alternatives.

Aquafaba (AQ) is the viscous liquid resulting from cooking chickpea seed or other legumes in water [8]. AQ has been gaining popularity since 2014, when a novel recipe blogger used the leftover liquid from a chickpea can as an egg replacement in vegan meringue [9]. Due to its desirable foaming and emulsification properties, AQ is now widely used by the vegan community as an egg replacement in many food products, such as mayonnaise, meringues and baked goods. Chickpea AQ components have been identified by Shim et al. (2018) [10]. Its application as a foaming agent has been reported in several studies [11,12]. However, the substances conferring AQ egg-similar emulsion properties have only been partially elucidated. Meanwhile, AQ qualities differ among diverse cooking conditions and legume genotypes. Therefore, chickpea cultivar selection and AQ process standardization are required to assure the quality of both AQ and AQ-based emulsions.

Based on previous studies and AQ functional properties, the central hypothesis of this research is that AQ emulsion properties not only differ among chickpea cultivars, but also have correlations with chickpea seed components and physicochemical properties. Therefore, the primary objective of this study is to prepare AQ from major chickpea cultivars and use this product to produce food oil emulsions then compare the properties of those emulsions. In addition, physicochemical properties and hydration kinetics of the different chickpea cultivars used in this study were determined to investigate possible correlations among these parameters and AQ emulsion properties.

2. Materials and Methods

2.1. Materials

Four Kabuli chickpea cultivars (CDC Leader, CDC Orion, CDC Luna and Amit) and one Desi chickpea cultivar (CDC Consul) were generously provided by Dr. Bunyamin Tar'an from the University of Saskatchewan, Crop Development Centre (CDC, Saskatoon, SK, Canada). Seed was randomly selected and manually cleaned and freed of broken seed, dust and other foreign materials. Canola oil (purity 100%; ACH Food Companies, Inc., Terrace, IL, USA) and baking soda ($NaHCO_3$; ARM & HAMMER by Church and Dwight Co., Inc., Mississauga, ON, Canada) were purchased from a local supermarket (Walmart, Saskatoon, SK, Canada). Sodium dodecyl sulfate (SDS) was purchased from GE Healthcare (Mississauga, ON, Canada). Anhydrous ether was obtained from Fisher Scientific Co. (Ottawa, ON, Canada). Sodium hydroxide (NaOH) and sodium chloride (NaCl) were purchased from Sigma-Aldrich Canada Ltd. (Oakville, ON, Canada). Concentrated sulphuric acid (H_2SO_4, ≥96%, *w/w*) and methanol were acquired from EMD Millipore Corporation (Burlington, MA, USA).

2.2. Fresh AQ Preparation

Chickpea seed (100 g) was washed and soaked in distilled water at a ratio of 1:4 (*w/w*), covered, and kept at 4 °C for 16 h [11]. Soaking water was then drained and discarded. Soaked chickpea seed (100 g) was rinsed with distilled water and then mixed with 100 mL distilled water in 250 mL sealed glass jars and cooked in a pressure cooker at 115–118 °C (an autogenic pressure range of 70–80 kPa) for 30 min. Subsequently, jars of cooked chickpeas were cooled by holding at room temperature for 24 h. Cooled AQ was drained from cooked chickpea seed using a stainless-steel strainer, then stored in a freezer (−18 °C). AQ samples from each chickpea cultivar were prepared in quadruplicate. Prior to analysis, AQ was thawed at 4 °C overnight then held at room temperature for 2 h. AQ moisture content was determined by oven drying at 105 °C overnight according to the American Association of Cereal Chemists (AACC) method 44-15.02 (AACC, 2000) [13].

2.3. AQ Emulsion Properties

2.3.1. AQ Oil Emulsion Preparation

Freshly thawed AQ (6 g), produced from each cultivar, was mixed with 14 g canola oil using a kitchen hand mixer. The mixer was set at its maximum speed for 2 min. Canola oil was added dropwise to the AQ to produce emulsions. The emulsion type (O/W or W/O) was determined by a simple dilution test: A small amount of emulsion was dispersed into two beakers, one containing the oil phase (canola oil) and the other containing the aqueous phase (water). An easy dispersion occurs only in the continuous phase of the emulsion [14]. All emulsions prepared in this study dispersed easily in water, and were thereby confirmed to be O/W emulsions.

2.3.2. Emulsion Capacity

Each AQ oil emulsion was diluted 100-fold with 0.1% SDS (*w/v*), and emulsion turbidity (500 nm) was calculated immediately after dilution. A UV-Vis spectrophotometer was used to determine transmittance at 500 nm. Emulsion turbidity value (T) was calculated using Equation (1).

$$T = \frac{2.303 \times A \times V}{I}, \tag{1}$$

in which T is the emulsion turbidity (m^{-1}), A is the emulsion "absorbance" measurement at 500 nm (1/transmittance), V is the dilution factor and I is the path length (0.01 m).

AQ emulsion capacity (EC) was determined according to Liu et al. (2016) [15]. An emulsifying activity index (EAI) was used as an indicator and defined by Wang et al. (2010) and Pearce and Kinsella (1978) using Equations (2) [16] and 3 [17], respectively.

$$EAI = \frac{2T}{\varnothing \times C}, \tag{2}$$

in which $\varnothing$ is the oil volume fraction of dispersed phase and C is the emulsifier concentration (the weight of AQ per unit volume of the aqueous phase before the emulsion is formed) [17].

$$\varnothing = \frac{C - A_1 - E(B - C)}{C - A_1 + \frac{(B-C)\{(1+E)\,D_0 - E\}}{D_s}}. \tag{3}$$

In Equation (3), A_1 denotes mass of beaker; B is mass of beaker plus emulsion; C is the mass of beaker plus emulsion dry matter; D_0 is the oil density; D_s is the protein solution density, and E is the solute concentration (mass per unit of solvent) [17]. All measurements were conducted in triplicate, and results expressed as mean ± standard deviation (SD).

2.3.3. Emulsion Stability

Emulsion stability (ES) value was determined at room temperature. Emulsions were transferred to sealed 15 mL centrifuge tubes, which were then centrifuged at 1860× *g* for 15 min. The weight of the original emulsion before centrifugation (F_0) and the emulsified layer (F_1) after centrifugation were measured. The emulsion stability at room temperature was determined by Equation (4) [18].

$$ES = \frac{F_1}{F_0} \times 100\%. \tag{4}$$

All measurements were conducted in triplicate, and results expressed as mean ± SD.

2.4. Chickpea Physical Properties

Hundred seed weight (*HSW*, g) was determined by randomly selecting and weighing 100 grains selected from each chickpea cultivar. Seed coat incidence (*SCI*, %) was determined by the method of Alova and Patanè (2010) with minor modification [19]. The seed coats of ten chickpea seeds were removed after soaking seed in distilled water at 4 °C for 12 h. Then the seed coat and cotyledons were dried separately at 65 °C for 4 h, and weighed each hour until a constant weight was recorded.

Seed dimensions were determined by randomly selecting ten chickpea seeds and then using a micrometer to record the seed dimensions in three perpendicular directions. Equation (5) was used to calculate the geometric average of the diameter of an equivalent dimension (*ED*, mm).

$$ED = (L \times W \times T)^{1/3},\tag{5}$$

in which L, W and T were the major, minor and intermediate axes (mm), respectively [20].

The surface area per unit mass of seed (or, specific surface area, *SSA*, mm^2/mg) and seed coat weight per surface area (namely seed coat thickness, *WSA*, mg/cm^2) of a single seed were calculated based on the *ED* and *HSW* value by the following Equations (6) and (7), respectively.

$$SSA = \frac{\pi \times ED^2 \times 100}{HSW},\tag{6}$$

$$WSA = \frac{HSW}{100 \times \pi \times ED^2}.\tag{7}$$

2.5. Chickpea Hydration Kinetics

Hydration kinetic tests were performed by the method of Avola and Patanè (2010) [19] with minor modification. Chickpea seed was soaked at room temperature (20–22 °C) and weighed periodically to determine water uptake kinetics. Ten seeds were transferred to a 200 mL beaker, which contained 150 mL deionized water or aqueous solutions of 0.5% (*w/v*) NaCl or NaHCO$_3$. Beakers were held at a constant temperature of 22 °C. Each hour up to the eighth hour, then at 24 h after initial imbibition, the seed was drained, and then weighed after free water was absorbed with a low-lint wiper. A clean wiper was used for each weighing to avoid contamination with solutes or water.

A two-parameter asymptotic Equation (8) was used to model water uptake kinetics (SigmaPlot 9.0; Systat Software, Inc., San Jose, CA, USA).

$$H_t = H_{max} \times \left(1 - e^{-kx}\right),\tag{8}$$

in which H_t is hydration weight (g/seed) after soaking for time t (h), H_{max} is the asymptote of the curve (to estimate seed weight at full hydration), k is a curve parameter that is related to the initial hydration rate (estimating H_{rate}). All measurements were conducted in triplicate, and results expressed as mean ± SD.

2.6. Chickpea Chemical Properties

The moisture content of whole chickpea seed was measured by the ASAE S352.2 air oven drying method (103 °C, 72 h, 15 g) [21]. Selected whole chickpea seed was ground with a disc mill before proximate composition analysis. Analyses of crude protein, crude fat, ash and crude fibre were performed using Association of Official Analytical Chemists (AOAC) methods [22]. In brief, nitrogen content was analyzed by combustion (AOAC Method 990.03) [22] using a LECO (Saint Joseph, MI, USA) nitrogen analyzer. Protein content was calculated as nitrogen content multiplied by a conversion factor 6.25. Fat was extracted from ground samples according to AOAC method 920.39 [22] using anhydrous ether in a Soxhlet apparatus (Extraction system B-811, BÜCHI Labortechnik AG., Switzerland). Ground chickpea samples were weighed (2 g) onto filter paper which was then placed in a cellulose Soxhlet extraction thimble and washed five times with 20 mL distilled H_2O each time.

After drying in an oven at 102 °C for 2 h, oil was extracted over 5 h in a Soxhlet apparatus with anhydrous ether. Chickpea ash content was determined by the AOAC method 942.05 [22]. Samples were weighed (2 g) in separate, pre-weighed porcelain crucibles, and placed in a preheated furnace (600 °C) for 2 h. Crucibles were then transferred to a desiccator, cooled and reweighed. Sample weight remaining after ignition of a 2 g sample was regarded as ash content. Crude fibre content was determined by AOAC method 962.09 with minor modification [22]. Samples were digested with 1.25% (*w/v*) boiling H_2SO_4 (30 min) followed by 1.25% (*w/v*) boiling NaOH (30 min) and washed with methanol. Samples were then dried to a constant weight and the residue burned. Weight loss on ignition of the dried residue was regarded as crude fibre content. Carbohydrate content was determined by subtracting the total percentage of protein, fat, fibre and ash components from 100%. All measurements were conducted in triplicate, and results were expressed as mean ± SD.

2.7. Statistical Analysis

Three replications were used to obtain the average and SD values for all tests. Data are presented as mean ± SD (*n* = 3). Analytical results were processed with Microsoft Excel 2018. Statistics were implemented through the Statistical Package for the Social Science (SPSS) version 25.0 (IBM Corp., Armonk, NY, USA). The analysis of variance (ANOVA) and Tukey's tests were used to evaluate the statistical significance of differences in properties and composition. Statistical significance was accepted at $p < 0.05$. The mathematical model parameters used in chickpea seed hydration kinetics measurements were estimated using a nonlinear regression procedure performed using SigmaPlot software (Systat Software Inc. San Jose, CA, USA). Model suitability was evaluated using the coefficient of determination (R^2), which indicates the model predictive quality (the higher the value for R^2, the better the goodness of fit, and up to a value of 1, meaning exact fit). The hydration kinetics parameters given by the nonlinear regressions were used to compare chickpea cultivars and soaking treatments, including soaking time and soaking solutions. Pearson correlation coefficients (*r*) for the relationships between all characteristics were calculated.

3. Results

3.1. AQ Produced from Different Chickpea Cultivars

AQ prepared from different chickpea cultivars showed significantly different yields and moisture contents (Figure 1). Liquid AQ yields ranged from 70.90 g/100 g seed to 107.44 g/100 g seed, with the highest yield produced by CDC Luna and the lowest by CDC Leader. AQ moisture content ranged from 92.4% to 94.2%, with the highest moisture content in AQ produced by CDC Luna and the lowest by CDC Leader. Commercially, high yield AQ with low moisture content (high dry matter content) would be of greater economic value.

Colour and turbidity of AQ varied with chickpea cultivar (Figure 2). In the current study, CDC Leader, CDC Orion, CDC Luna and Amit are Kabuli class chickpea cultivars which normally have a white to cream–yellow colour seed, while CDC Consul is a Desi class chickpea with brown to fawn colours. AQ produced from CDC Leader and CDC Orion had similar colour and high turbidity. The AQ samples were pale yellow and cloudy liquids. Whereas, AQ produced from Amit was bright yellow and cloudy. Interestingly, AQ produced by CDC Luna had the lowest turbidity, and was nearly translucent and bright yellow. Also remarkable, AQ produced by CDC Consul was a dark brown colour, and had high turbidity. This dark brown colour might arise from tannins in the CDC Consul seed coat that may have migrated to the water during cooking [23]. AQ colour is also influenced by other water-soluble molecules in chickpea seed such as pigments, vitamins and other plant secondary metabolites. Chickpea contains pigments mainly falling into the carotenoids class (β-carotene, cryptoxanthin, lutein and zeaxanthin) as well as small amounts of chlorophyll [24]. Moreover, there are also water-soluble vitamins in chickpea, such as thiamin, riboflavin, niacin, vitamin B_6, folate and ascorbic acid [25]. Additionally, some flavonoid compounds, including anthocyanin, flavonols, isoflavones, flavonol glucosides, phlobaphenes, proanthocyanidin, leucoanthocyanidin and proanthocyanidin in the seed coat might contribute to AQ colour [26].

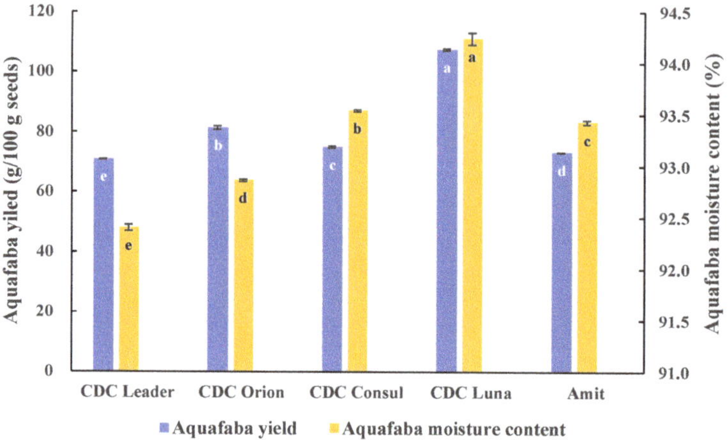

Figure 1. Fresh Aquafaba (AQ) yield (g/100 g seed) and moisture content (%) prepared from different chickpea cultivars from Crop Development Centre (CDC) in Saskatoon, SK, Canada. Means within the same property without a common letter (a–e) are significantly different according to Tukey's test.

Figure 2. (**A**) AQ and seed of different chickpea cultivars prepared in jars. From left to right: CDC Consul, CDC Luna, CDC Orion, CDC Leader, and Amit; (**B**) AQ separated from chickpea seed. From left to right: CDC Consul, CDC Luna, CDC Leader, CDC Orion, and Amit.

In general, hemicellulose [27] and cellulose [28] are probably disrupted by cooking for 30 min at 115–118 °C and autogenic pressure (70–80 kPa), leading to partial destruction of the chickpea cell wall and breaking bonds between lignin and hemicelluloses [29]. Therefore, AQ turbidity and colour is a result of the disruption of chickpea seed microstructure during cooking, leaching of organic compounds, pigments, proteins, sugars, starch and vitamins into the cooking water [30].

3.2. AQ Emulsification Properties

3.2.1. AQ Emulsion Capacity

Today, healthier and nutritious foods are demanded by health-conscious consumers. Food oil-in-water emulsions, such as mayonnaise and salad dressing, are often avoided due to their high fat and cholesterol content. Plant-based protein fractions, including soybean and wheat proteins, are ingredients that may be used in replacing egg as emulsifiers in mayonnaise emulsion systems [31,32]. Nikzade et al. (2012) developed a combination of soy milk, gums and other stabilizers to replace egg in low cholesterol-low fat mayonnaise formulas [18]. The application and development of different ingredients in reduced fat/cholesterol salad dressing and mayonnaise have been summarized by Ma and Boye (2013) [33].

EAI values of AQ prepared from each of five chickpea cultivars were measured (Figure 3). *EAI* ranged from 1.10 ± 0.04 to 1.30 ± 0.05 m^2/g with the highest *EAI* observed for AQ prepared from CDC Leader (1.30 ± 0.05 m^2/g), while the lowest *EAI* occurred with CDC Orion (1.10 ± 0.04 m^2/g). The *EAI*s of CDC Consul (1.21 ± 0.02 m^2/g), CDC Luna (1.17 ± 0.07 m^2/g), and Amit (1.20 ± 0.05 m^2/g) were not statistically different from each other.

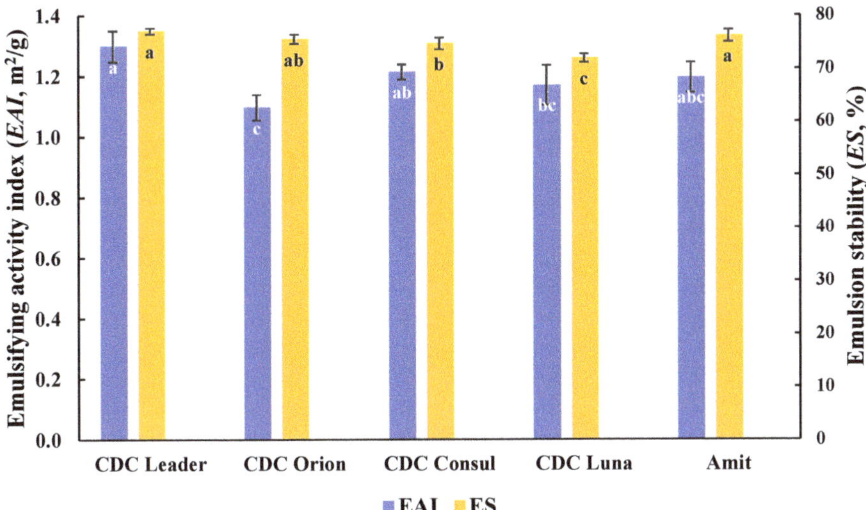

Figure 3. Emulsion capacity and stability of AQ prepared from different chickpea cultivars. Means within the same property without a common letter (a–c) are significantly different according to Tukey's test.

3.2.2. AQ Emulsion Stability

Emulsion stability of AQ from cooked whole seed of five chickpea cultivars was also investigated in this study (Figure 3). AQ emulsion stability ranged from 71.9 ± 0.8% to 77.1 ± 0.5%, with the highest emulsion stability also observed from AQ prepared by CDC Leader and Amit, while the lowest emulsion stability for AQ prepared by CDC Luna. The emulsion stability of AQ from CDC Orion (75.6%) and CDC Consul (74.7%) were not statistically different.

As a novel water-soluble emulsifier, AQ can stabilize oil-in-water emulsions to prepare egg-free vegan food oil emulsions. Additionally, our results showed that AQ prepared from Kabuli type (CDC Leader) chickpea exerted the best emulsion capacity and stability compared to the other chickpea cultivars studied in this research. This indicates the potential for selecting CDC Leader to produce an AQ emulsifier. This difference in emulsification properties between different chickpea cultivars is probably related to the difference in the physical properties and chemical compositions of chickpea seed which affect the mass transfer to the cooking water during cooking.

To better understand the emulsion properties differences of AQ prepared from different chickpea cultivars, we studied the physiochemical properties of chickpea seed, and we tested for correlations between these properties and AQ emulsion properties.

3.3. Chickpea Physical Properties

Seed coat cracking after soaking and during cooking results from splitting of the outer cell wall layers. During chickpea cooking, this seed coat works as a membrane that controls mass transfer, which would affect the composition, and therefore, the functional properties of the resulting cooking water (AQ) [8]. Seed coat physical characteristics depend on the genotype and environmental conditions (temperature, soil and moisture) at the time of seed maturity or during storage. Physical characteristics of seed from different chickpea cultivars are shown in Table 1. Significant differences were observed in *HSW*, *ED*, *SCI* and *WSA* among chickpea cultivars. *HSW* and *ED* ranged from 22.43 ± 0.08 g to 42.9 ± 0.3 g and 6.89 ± 0.5 mm to 8.49 ± 0.2 mm, respectively. CDC Leader exhibited heavier (42.90 ± 0.3 g/100 seed) and larger (8.49 ± 0.2 mm) seed compared with the other cultivars.

CDC Orion was not significantly ($p > 0.05$) different from CDC Luna, except for *WSA* (10 ± 1 mg/cm^2). CDC Consul exhibited the greatest *SCI* (11.2 ± 2%) and *WSA* (15 ± 1 mg/cm^2). Amit showed the highest *SSA* (0.668 ± 0.09 mm^2/mg). These observed values are comparable to results reported previously [19], where three Sicilian chickpea cultivars (Calia, Etna and Principe) were evaluated for their *HSW*, *ED*, *SCI*, *SSA* and *WSA*. In their study, the chickpea *HSW* value ranged from 31.3 g/100 seed to 48.8 g/100 seed. Three Sicilian chickpea cultivars had similar *ED* and *SCI* with an average of 7.8 mm and 5.78%, respectively. The *SSA* value differed among chickpea cultivars and ranged from 0.45 mm^2/mg to 0.6 mm^2/mg. The *WSA* value showed a wide range from 8.3 to 11.9 mg/cm^2. These differences in the physical characteristics between different chickpea cultivars will be reflected in the seed coat behavior during soaking and cooking and might explain the variation in AQ properties. CDC Leader exhibited a good seed weight (42.90 ± 0.3 g) with the lowest *SCI* (3.89 ± 0.3%) which might explain why AQ prepared from this variety has the highest dry material (7.6%) and emulsion capacity and stability. *SCI* reflects fiber content, seed coat thickness and compactness, which is correlated with the diffusion resistance and leaching of soluble solids during soaking and cooking. The differences in the cookability of different chickpea genotypes have been reported previously which they attributed to the difference in the seed characteristics [8,19].

Table 1. Physical and chemical characteristics of seed from five chickpea cultivars.

Characteristics	Unit	CDC Leader	CDC Orion	CDC Consul	CDC Luna	Amit
Physical						
Hundred seed weight, *HSW*	g	42.90 ± 0.3 [a]	41.03 ± 0.6 [b]	33.34 ± 0.4 [d]	40.37 ± 0.6 [bc]	22.43 ± 0.08 [e]
Equivalent dimension, *ED*	mm	8.49 ± 0.2 [a]	8.45 ± 0.3 [a]	8.21 ± 0.2 [a]	8.24 ± 0.1 [a]	6.89 ± 0.5 [b]
Seed coat incidence, *SCI*	%	3.89 ± 0.3 [c]	6.63 ± 2 [b]	11.2 ± 2 [a]	5.39 ± 0.6 [bc]	4.65 ± 0.9 [c]
Specific surface area, *SSA*	mm^2/mg	0.528 ± 0.03 [b]	0.547 ± 0.04 [b]	0.621 ± 0.03 [ab]	0.529 ± 0.02 [b]	0.668 ± 0.09 [a]
Seed coat weight per surface area, *WSA*	mg/cm^2	6.9 ± 0.4 [d]	10 ± 1 [c]	15 ± 1 [a]	5.7 ± 0.2 [d]	12 ± 2 [b]
Technological						
Hydration capacity (t = ∞), H_{max}	g (H_2O g/dw)	1.036 ± 0.02 [b]	1.073 ± 0.01 [b]	0.9870 ± 0.03 [b]	1.025 ± 0.1 [b]	1.198 ± 0.03 [a]
Hydration rate, H_{rate}	g (H_2O g/min)	0.3537 ± 0.01 [ab]	0.3914 ± 0.04 [ab]	0.3412 ± 0.09 [b]	0.4542 ± 0.03 [a]	0.4112 ± 0.04 [ab]
Chemical						
Moisture	%	8.86 ± 0.07 [c]	9.24 ± 0.08 [b]	10.7 ± 0.1 [a]	5.48 ± 0.02 [d]	5.29 ± 0.01 [e]
Carbohydrate	g (100 g/dw)	65.4 ± 2 [ab]	61.8 ± 2 [b]	65.2 ± 2 [a]	67.4 ± 0.7 [a]	66.8 ± 1 [ab]
Protein	g (100 g/dw)	20.9 ± 0.1 [b]	23.6 ± 0.08 [a]	18.7 ± 1 [c]	18.3 ± 0.3 [c]	20.2 ± 0.1 [b]
Fat	g (100 g/dw)	6.49 ± 0.5 [ab]	5.96 ± 0.5 [b]	4.64 ± 0.5 [c]	7.24 ± 0.5 [a]	4.10 ± 0.5 [cd]
Ash	g (100 g/dw)	3.0 ± 0.1 [c]	3.2 ± 0.0 [b]	2.9 ± 0.1 [d]	2.5 ± 0.0 [e]	3.4 ± 0.1 [a]
Fibre	g (100 g/dw)	4.32 ± 1 [b]	5.36 ± 2 [ab]	8.59 ± 0.6 [a]	4.63 ± 1 [b]	5.55 ± 0.8 [ab]

Data are expressed as mean ± SD (n = 3). Value within rows followed by the same letter (e.g., [a,b,c,d]) indicates no significant difference ($p > 0.05$) between varieties by Tukey's test. CDC: Crop Development Centre (Saskatoon, SK, Canada).

3.4. Hydration Kinetics

Water absorption capacity during soaking is generally related to the physical properties of the seed. Hydration of the testa and swelling of the cotyledons soften the cell walls and change tissue permeability, reduce the cooking time, and affect mass transfer to the cooking water. The relationship between soaking time and cumulative values of water uptake (Figure 4) was described by a nonlinear iterative regression method with an exponential relationship (Equation (8)). The applied model fitted the experimental data with an R^2 that was greater than 0.94 for all cultivars. Therefore, a single curve for water uptake was used in further analysis with all data combined. Soaking processes achieve rapid water uptake (H_{rate} = 0.38 g H_2O g/min), and after soaking for 6 h, water absorbed reached 90% of seed dry weight. Subsequently, the water absorption rate declined until the hydrated seed weight was 2.06-fold greater than before hydration, where total hydration reached saturation at 1.06 g H_2O g/dw (H_{max}). Similar trends for chickpea seed hydration were described by several authors [19,34,35]. However, water content of chickpea seed exceeded 90% of total water imbibition after 4 h [19].

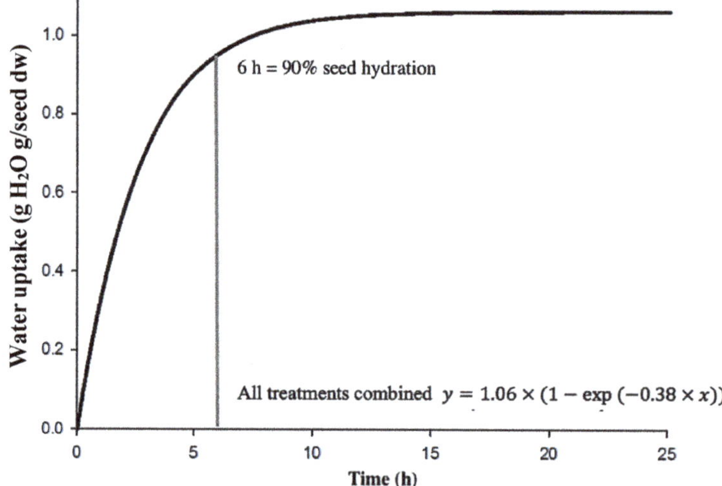

Figure 4. Chickpea seed water absorption kinetics. A common curve fitted all data (different chickpea cultivars and hydration solutions).

Statistical analysis revealed no significant differences among H_{max}, except Amit seed which absorbed more water than the other chickpea cultivars (1.20 g H_2O g/dw) in all soaking solutions (Tables 1 and 2). The modeled hydration rate (H_{rate}) of these five chickpea cultivars were mostly similar and ranged from 0.328 to 0.417 g H_2O g/dw. Interestingly, the H_{max} and H_{rate} of chickpea seed soaking in NaCl solution was slightly lower compared with that obtained in deionized water. However, seed of CDC Leader, exhibited similar H_{rate} in both deionized water and NaCl solution. Conversely, soaking seed in NaHCO$_3$ solution increased H_{max} for CDC Luna and H_{rate} for CDC Consul. These results are partially in agreement with previous reports that the presence of salt in soaking solution results in slowed seed hydration [36], but contrast with the findings of Avola and Patanè (2010) and Clemente et al. (1998), where no effect was observed on the hydration of chickpea seed soaking in salt solutions [19,34].

There were two possible explanations for increased H_{max} results in NaHCO$_3$ solution: (1) the osmotic pressure gradient across membranes of cotyledon cells was decreased [37], (2) there was an interaction of carbonate ions with biopolymers in cotyledon cells which might produce molecular unfolding with a possible exposure of new sites for water binding [38].

Table 2. Kinetic constants of the nonlinear regression analysis for chickpea seed hydration.

Hydration Solution	H_{max} g H_2O g/dw	H_{rate} g/min	R^2
CDC Leader			
H_2O	1.058	0.356	0.996
NaCl	1.014	0.359	0.996
$NaHCO_3$	1.035	0.347	0.996
CDC Orion			
H_2O	1.078	0.431	0.992
NaCl	1.060	0.364	0.993
$NaHCO_3$	1.080	0.379	0.992
CDC Consul			
H_2O	1.023	0.313	0.974
NaCl	0.973	0.268	0.941
$NaHCO_3$	0.965	0.442	0.968
CDC Luna			
H_2O	0.993	0.479	0.986
NaCl	0.948	0.425	0.993
$NaHCO_3$	1.133	0.459	0.943
Amit			
H_2O	1.192	0.456	0.988
NaCl	1.172	0.384	0.998
$NaHCO_3$	1.230	0.393	0.988
All data combined	1.062	0.380	0.929

H_{max}, Max hydration capacity; H_{rate}, initial hydration rate.

3.5. Chickpea Chemical Properties

The main chemical constituents (moisture, carbohydrate, protein, fat, ash and fibre content) of different cultivars of chickpea are summarized in Table 1. The moisture content of raw chickpea seed showed significant difference among chickpea cultivars (5.29 ± 0.01% to 10.7 ± 0.1%), with the highest moisture content for CDC Consul and the lowest for Amit. Carbohydrate was the main component in all of the samples, while protein was the second most abundant component. CDC Luna had the highest carbohydrate content (67.4 ± 0.7 g 100 g/dw), followed by Amit (66.8 ± 1 g 100 g/dw) and CDC Leader (65.4 ± 2 g 100 g/dw). It is important to note that CDC Orion had the lowest carbohydrate content, which might be correlated with the lowest emulsion properties observed for AQ prepared from this cultivar.

Protein content ranged from 18.3 ± 0.3 (CDC Luna) to 23.6 ± 0.08 g 100 g/dw (CDC Orion). The former also contained more fat (7.24 g ± 0.5 g 100 g/dw) than other chickpea cultivars, while Amit had the lowest fat content (4.10 ± 0.5 g 100 g/dw). Chickpea ash content did not differ with genotype. The mean value of ash content was 3.0 g 100 g/dw. Crude fibre content ranged from 4.32 ± 1 g 100 g/dw to 8.59 ± 0.6 g 100 g/dw. These observations are in agreement with previous studies by Xu et al. (2014), Özer et al. (2010), and de Almeida Costa et al. (2006) for chemical composition of raw chickpea seed from different chickpea cultivars [39–41]. In addition, Khattak et al. (2006) analyzed protein and ash content of seven Kabuli chickpea cultivars, which ranged from 18.08 to 19.22% and 2.45% to 2.94% [42], and thus, was similar to the values in this study.

3.6. Correlation Analysis

The overall interrelationships among chickpea physical, proximate composition (protein, fat, carbohydrate, fibre and ash), hydration characteristics, and AQ yield, emulsion capacity and stability are shown in Table 3.

Table 3. Correlation coefficient among the sixteen physical, chemical, and hydration attributes for the five chickpea cultivars and AQ yield, emulsion capacity, and emulsion stability.

	ES	EC	HSW	SCI	ED	SSA	WSA	Carboh.	AQ moisture	Protein	Fibre	Fat	Ash	H_{max}	H_{rate}
AQ yield	−0.94 *	−0.28	0.47	0.09	0.42	−0.55	−0.50	−0.29	0.71	−0.48	−0.15	0.71	−0.92 *	−0.54	0.32
ES		0.40	−0.17	−0.24	−0.16	−0.24	0.25	−0.37	−0.91 *	−0.58	−0.06	−0.44	0.82	0.39	−0.37
EC			0.02	0.20	0.01	−0.00	−0.02	0.47	−0.34	0.39	−0.06	−0.01	−0.08	−0.17	−0.51
HSW				−0.10	0.97 **	−0.97 **	−0.63	−0.39	−0.28	0.24	−0.38	0.87	−0.52	−0.71	−0.19
SCI					0.15	0.30	0.75	0.24	0.26	−0.26	0.95 *	−0.39	−0.18	−0.52	−0.62
ED						−0.88	−0.42	−0.48	−0.28	0.21	−0.13	0.74	−0.51	−0.83	−0.38
SSA							0.80	0.25	0.19	−0.21	0.57	−0.96 *	0.55	0.58	−0.03
WSA								−0.20	0.02	−0.04	0.90 *	−0.90 *	0.45	0.12	−0.47
Carbohydrate									0.59	−0.85	−0.11	−0.01	−0.40	−0.14	0.27
AQ moisture										−0.72	0.22	0.04	−0.59	−0.07	0.43
Protein											−0.31	0.01	0.65	0.30	0.09
Fibre												−0.64	0.02	−0.30	−0.63
Fat													−0.68	−0.51	0.23
Ash														0.74	0.01
H_{max}															0.60

AQ, aquafaba; ES, emulsion stability; EC, emulsion capacity; HSW, 100 seed weight; ED, equivalent dimension; SCI, seed coat incidence; SSA, specific surface area; WSA, weight of seed coat per surface area; H_{max}, max hydration capacity; H_{rate}, initial hydration rate. *, ** indicate significant for $p < 0.05$ and 0.01, respectively.

AQ yield was inversely correlated to both *ES* ($r = -0.94^*$) and ash content ($r = -0.92^*$). Emulsion stability was also negatively correlated to AQ moisture content ($r = -0.91^*$), suggesting that AQ with higher dry matter contents have better emulsion properties. Seed coat incidence, *SSA* and *WSA* were not related to seed dimension, but *SSA* was found to have the highest negative correlation with seed weight ($r = -0.97^{**}$). *HSW* was also positively correlated to *ED* ($r = 0.97^{**}$), indicating that heavier and larger chickpea seeds develop a smaller surface area per unit mass. Fibre was closely correlated to *SCI* ($r = 0.95^*$) and seed coat thickness (*WSA*: $r = 0.90^*$). Therefore, a chickpea cultivar with low *SCI* and *WSA* will have low fibre content in the seed coat [43]. These results were observed for CDC Leader, which had the lowest *SCI* and produced AQ with superior emulsion formation and stabilization properties.

Fat content was negatively correlated with *SSA* ($r = -0.96^*$) and *WSA* ($r = -0.90^*$), complementary to Gil et al. (1996) who observed a similar relationship in Desi and Kabuli chickpea [44]. In their study, fat content was also positively, and significantly, correlated to *HSW* for both chickpea classes. However, in this study, correlation between chickpea fat content and *HSW* ($r = 0.87$) was insignificant. This study supported observations made by Khattak et al. (2006), who revealed a strong, positive correlation between seed size and seed weight [42]. Moreover, they found that seed size was positively correlated with chickpea hydration capacity and protein content, as well as moisture content. However, no similar correlation was observed in this study.

Chickpea emulsion properties have often been linked to carbohydrate content, protein content [45,46] and some phytochemicals, such as phenolics [47] and saponins [48,49]. Shim et al. [10] investigated AQ compositions recovered from commercially canned chickpea products and identified proteins present in AQ. They demonstrated that most proteins in AQ are mostly of small molecular weight (≤ 16.7 kDa) and many belong to the groups that include late embryogenesis abundant proteins (LEAP), dehydrins and defensin. Main carbohydrate types in AQ are simple sugars (such as sucrose and glucose), soluble and insoluble fiber including cellulose and pectin [11,12]. Importantly, the contribution of hydrophobic polysaccharides and amphiphilic phytochemicals to emulsification activity cannot be neglected. Improved rheological properties of hydrophilic polysaccharides induced steric and mechanical stabilization effects, which slowed or even prevented emulsion droplet aggregation by forming thick charged layers [50].

AQ also contains saponins, which are regarded as surfactants and emulsifiers due to their amphiphilic structure [11,49]. Chung et al. (2017) presumed that saponins could pack tightly together at the oil water interface, and thereby, effectively avoid unfavourable molecular interactions between the phases [51]. This could lower the interfacial tension to generate smaller droplets during homogenization and lead to higher emulsion stability [51].

Chickpea seed physicochemical properties and hydration characteristics did not correlate with AQ emulsion capacity and stability. This could be due to other factors that control the dispersal of chemical substances into AQ during cooking and the interaction between these components during cooking and storage.

4. Conclusions

This study assessed the emulsion capacity and stability of AQ produced from five chickpea cultivars grown in Canada, and investigated the overall interrelationship between AQ yield and emulsion properties and chickpea physical, chemical and hydration characteristics. Our results showed that AQ produced from different chickpea cultivars illustrated significant differences in emulsion capacity and stability. AQ emulsion capacity and stability, among the five chickpea cultivars, ranged from 1.10 to 1.30 m^2/g and 71.9 to 77.1%, respectively. Furthermore, AQ obtained from CDC Leader produced emulsions with superior emulsion capacity and stability compared to AQ prepared from other chickpea cultivars. In our study, we did not observe a significant correlation between the proximate composition of chickpea seed (carbohydrate, protein, etc.) and AQ emulsion properties (emulsion capacity and stability). However, this weak correlation did not suggest that AQ chemical components were not correlated to AQ emulsion properties, and vice versa.

AQ composition is a complex mixture of components transferred from seed to water during cooking, in addition to other molecules produced from the interaction between these components under high pressure and temperature. Therefore, further study is needed to determine the content of other components in chickpea seed and AQ, such as saponin, fibre content and type, Maillard reaction products, and to explain the variation in AQ properties of different chickpea cultivars. In conclusion, AQ exhibits excellent emulsification properties, and is a promising emulsifier that has a great potential to be used in preparation of novel, egg-free and vegan emulsion products such as mayonnaise and salad dressing. However, the selection of chickpea genotype is required to standardize AQ production and emulsion properties.

Author Contributions: Conceptualization, R.M., M.J.T.R.; validation, Y.Y.S. and V.M.; investigation, Y.H.; writing-original draft preparation, Y.H.; writing-review and editing, Y.Y.S., R.M., and M.J.T.R.; supervision, V.M. and M.J.T.R.; funding acquisition, M.J.T.R.

Funding: This research was supported by the Saskatchewan Ministry of Agriculture's Agriculture Development Fund (20180091) and by the Department of Chemical and Biological Engineering, BLE Devolved Scholarship (University of Saskatchewan, Canada).

Acknowledgments: In this section you can acknowledge any support given which is not covered by the author contribution or funding sections. This may include administrative and technical support, or donations in kind (e.g., materials used for experiments).

Conflicts of Interest: The authors declare no conflict of interest.

References

1. Baldwin, R.E. Functional properties of eggs in foods. In *Egg Science and Technology*; Stadelman, W.J., Cotterill, O.J., Eds.; AVI Publishing Company Inc.: Westport, CT, USA, 1977; pp. 279–290.
2. Moros, J.E.; Franco, J.M.; Gallegos, C. Rheological properties of cholesterol-reduced, yolk-stabilized mayonnaise. *J. Am. Oil Chem. Soc.* **2002**, *79*, 837–843. [CrossRef]
3. Laca, A.; Sáenz, M.C.; Paredes, B.; Díaz, M. Rheological properties, stability and sensory evaluation of low-cholesterol mayonnaises prepared using egg yolk granules as emulsifying agent. *J. Food Eng.* **2010**, *97*, 243–252. [CrossRef]
4. Park, H.Y.; Yoon, T.J.; Kim, H.H.; Han, Y.S.; Choi, H.D. Changes in the antigenicity and allergenicity of ovalbumin in chicken egg white by *N*-acetylglucosaminidase. *Food Chem.* **2017**, *217*, 342–345. [CrossRef]
5. Duan, X.; Li, M.; Ji, B.; Liu, X.; Xu, X. Effect of fertilization on structural and molecular characteristics of hen egg ovalbumin. *Food Chem.* **2017**, *221*, 1340–1345. [CrossRef]
6. Lin, M.; Tay, S.H.; Yang, H.; Yang, B.; Li, H. Replacement of eggs with soybean protein isolates and polysaccharides to prepare yellow cakes suitable for vegetarians. *Food Chem.* **2017**, *229*, 663–673. [CrossRef]
7. Janssen, M.; Busch, C.; Rödiger, M.; Hamm, U. Motives of consumers following a vegan diet and their attitudes towards animal agriculture. *Appetite* **2016**, *105*, 643–651. [CrossRef]
8. Mustafa, R.; Reaney, M.J.T. Chapter 4. Aquafaba, from food waste to a value-added product. In *Food Wastes and by-Products: Nutraceutical & Health Potential*; Campos-Vega, R., Oomah, B.D., Vergara-Castañeda, H.A., Eds.; John Wiley & Sons Ltd.: New York, NY, USA, 2020; pp. 93–126.
9. Révolution végétale. Mousses–Isolats de Protéines. Available online: http://www.revolutionvegetale.com/fr/non-classe/mousses-isolats-de-proteines/ (accessed on 29 November 2019).
10. Shim, Y.Y.; Mustafa, R.; Shen, J.; Ratanapariyanuch, K.; Reaney, M.J.T. Composition and properties of aquafaba: Water recovered from commercially canned chickpeas. *J. Vis. Exp.* **2018**, *132*, e56305. [CrossRef]
11. Stantiall, S.E.; Dale, K.J.; Calizo, F.S.; Serventi, L. Application of pulses cooking water as functional ingredients: The foaming and gelling abilities. *Eur. Food Res. Technol.* **2018**, *244*, 97–104. [CrossRef]
12. Mustafa, R.; He, Y.; Shim, Y.Y.; Reaney, M.J.T. Aquafaba, wastewater from chickpea canning, functions as an egg replacer in sponge cake. *Int. J. Food Sci. Tech.* **2018**, *53*, 2247–2255. [CrossRef]
13. AACC International. Moisture-Air Oven Methods, drying at 103 °C. Methods 44-15.02. In *Approved Methods of Analysis*, 11th ed.; Paul, St., Ed.; American Association of Cereal Chemists: Washington, MN, USA, 2000.

14. Zaouadi, N.; Cheknane, B.; Hadj-Sadok, A.; Canselier, J.P.; Hadj Ziane, A. Formulation and optimization by experimental design of low-fat mayonnaise based on soy lecithin and whey. *J. Dispers. Sci. Technol.* **2015**, *36*, 94–102. [CrossRef]

15. Liu, J.; Shim, Y.Y.; Shen, J.; Wang, Y.; Ghosh, S.; Reaney, M.J.T. Variation of composition and functional properties of gum from six Canadian flaxseed (*Linum usitatissimum* L.) cultivars. *Int. J. Food Sci. Technol.* **2016**, *51*, 2313–2326. [CrossRef]

16. Wang, Y.; Li, D.; Wang, L.J.; Li, S.J.; Adhikari, B. Effects of drying methods on the functional properties of flaxseed gum powders. *Carbohydr. Polym.* **2010**, *81*, 128–133. [CrossRef]

17. Pearce, K.N.; Kinsella, J.E. Emulsifying properties of proteins: Evaluation of a turbidimetric technique. *J. Agric. Food Chem.* **1978**, *26*, 716–723. [CrossRef]

18. Nikzade, V.; Tehrani, M.M.; Saadatmand-Tarzjan, M. Optimization of low-cholesterol-low-fat mayonnaise formulation: Effect of using soy milk and some stabilizer by a mixture design approach. *Food Hydrocoll.* **2012**, *28*, 344–352. [CrossRef]

19. Avola, G.; Patanè, C. Variation among physical, chemical and technological properties in three Sicilian cultivars of Chickpea (*Cicer arietinum* L.). *Int. J. Food Sci. Technol.* **2010**, *45*, 2565–2572. [CrossRef]

20. Mohsenin, N.N. *Physical Properties of Plant. and Animal Materials*; Gordon and Breach Science Publishers: New York, NY, USA, 1970.

21. ASABE Standards. *Moisture Measurement-Unground Grain and Seeds*; American Society of Agricultural and Biological Engineers: St. Joseph, MI, USA, 1988.

22. AOAC. *Official Methods of Analysis of Association of Official Analytical Chemists International*, 20th ed.; Horwitz, W., Latimer, G.W., Eds.; AOAC International: Gaithersburg, MD, USA, 2016; methods 990.03, 920.39, 942.05, and 962.09.

23. Lyimo, M.; Mugula, J.; Elias, T. Nutritive composition of broth from selected bean varieties cooked for various periods. *J. Sci. Food Agric.* **1992**, *58*, 535–539. [CrossRef]

24. Abbo, S.; Molina, C.; Jungmann, R.; Grusak, M.A.; Berkovitch, Z.; Reifen, R.; Kahl, G.; Winter, P.; Reifen, R. Quantitative trait loci governing carotenoid concentration and weight in seeds of chickpea (*Cicer arietinum* L.). *Theor. Appl. Genet.* **2005**, *111*, 185–195. [CrossRef]

25. USDA. Nutrient Database for standard Reference. Available online: http://ndb.nal.usda.gov/ndb (accessed on 22 November 2019).

26. Furukawa, T.; Maekawa, M.; Oki, T.; Suda, I.; Iida, S.; Shimada, H.; Takamure, I.; Kadowaki, K. The Rc and Rd genes are involved in proanthocyanidin synthesis in rice pericarp. *Plant J.* **2006**, *49*, 91–102. [CrossRef]

27. Hromádková, Z.; Ebringerová, A. Ultrasonic extraction of plant materials-investigation of hemicellulose release from buckwheat hulls. *Ultrason. Sonochem.* **2003**, *10*, 127–133. [CrossRef]

28. Sun, J.X.; Sun, X.F.; Zhao, H.; Sun, R. Isolation and characterization of cellulose from sugarcane bagasse. *Polym. Degrad. Stab.* **2004**, *84*, 331–339. [CrossRef]

29. Sun, J.X.; Sun, R.; Sun, X.F.; Su, Y. Fractional and physicochemical characterization of hemicelluloses from ultrasonic irradiated sugarcane bagasse. *Carbohydr. Res.* **2004**, *339*, 291–300. [CrossRef] [PubMed]

30. Yildirim, A.; Öner, M.D. Electrical conductivity, water absorption, leaching, and color change of chickpea (*Cicer arietinum* L.) during soaking with ultrasound treatment. *Int. J. Food Prop.* **2015**, *18*, 1359–1372. [CrossRef]

31. Ghoush, M.A.; Samhouri, M.; Al-Holy, M.; Herald, T. Formulation and fuzzy modeling of emulsion stability and viscosity of a gum–protein emulsifier in a model mayonnaise system. *J. Food Eng.* **2008**, *84*, 348–357. [CrossRef]

32. Puppo, M.C.; Sorgentini, D.A.; Añón, M.C. Rheological study of dispersions prepared with modified soybean protein isolates. *J. AM. Oil Chem. Soc.* **2000**, *77*, 63–71. [CrossRef]

33. Ma, Z.; Boye, J.I. Advances in the design and production of reduced-fat and reduced-cholesterol salad dressing and mayonnaise: A review. *Food Bioprocess Technol.* **2013**, *6*, 648–670. [CrossRef]

34. Clemente, A.; Sánchez-Vioque, R.; Vioque, J.; Bautista, J.; Millán, F. Effect of processing on water absorption and softening kinetics in chickpea (*Cicer arietinum* L.) seeds. *J. Sci. Food Agric.* **1998**, *78*, 169–174. [CrossRef]

35. Gowen, A.; Abu-Ghannam, N.; Frias, J.; Oliveira, J. Modelling the water absorption process in chickpeas (*Cicer arietinum* L.)-The effect of blanching pre-treatment on water intake and texture kinetics. *J. Food Eng.* **2007**, *78*, 810–819. [CrossRef]

36. Pinto, G.; Esin, A. Kinetics of the osmotic hydration of chickpeas. *J. Chem. Educ.* **2004**, *81*, 532–536. [CrossRef]

37. Woodstock, L.W. Seed imbibition: A critical period for successful germination. *J. Seed Technol.* **1988**, *12*, 1–15.
38. Leopold, A.C. Volumetric components of seed imbibition. *Plant Physiol.* **1983**, *73*, 677–680. [CrossRef]
39. Xu, Y.; Thomas, M.; Bhardwaj, H.L. Chemical composition, functional properties and microstructural characteristics of three Kabuli chickpea (*Cicer arietinum* L.) as affected by different cooking methods. *Int. J. Food Sci. Technol.* **2014**, *49*, 1215–1223. [CrossRef]
40. Özer, S.; Karaköy, T.; Toklu, F.; Baloch, F.S.; Kilian, B.; Özkan, H. Nutritional and physicochemical variation in Turkish Kabuli chickpea (*Cicer arietinum* L.) landraces. *Euphytica* **2010**, *175*, 237–249.
41. de Almeida Costa, G.E.; da Silva Queiroz-Monici, K.; Pissini Machado Reis, S.M.; de Oliveira, A.C. Chemical composition, dietary fibre and resistant starch contents of raw and cooked pea, common bean, chickpea and lentil legumes. *Food Chem.* **2006**, *94*, 327–330. [CrossRef]
42. Khattak, A.B.; Khattak, G.S.S.; Mahmood, Z.; Bibi, N.; Ihsanullah, I. Study of selected quality and agronomic characteristics and their interrelationship in Kabuli-type chickpea genotypes (*Cicer arietinum* L.). *Int. J. Food Sci. Technol.* **2006**, *41*, 1–5. [CrossRef]
43. Sreerama, Y.N.; Neelam, D.A.; Sashikala, V.B.; Pratape, V.M. Distribution of nutrients and antinutrients in milled fractions of chickpea and horse gram: Seed coat phenolics and their distinct modes of enzyme inhibition. *J. Agric. Food Chem.* **2010**, *58*, 4322–4330. [CrossRef]
44. Gil, J.; Nadal, S.; Luna, D.; Moreno, M.T.; Haro, A.D. Variability of some physico-chemical characters in Desi and Kabuli chickpea types. *J. Sci. Food Agric.* **1996**, *71*, 179–184. [CrossRef]
45. Karaca, A.C.; Low, N.; Nickerson, M. Emulsifying properties of chickpea, faba bean, lentil and pea proteins produced by isoelectric precipitation and salt extraction. *Food Res. Int.* **2011**, *44*, 2742–2750. [CrossRef]
46. Sila, A.; Bayar, N.; Ghazala, I.; Bougatef, A.; Ellouz-Ghorbel, R.; Ellouz-Chaabouni, S. Water-soluble polysaccharides from agro-industrial by-products: Functional and biological properties. *Int. J. Biol. Macromol.* **2014**, *69*, 236–243. [CrossRef]
47. Luo, Z.; Murray, B.S.; Yusoff, A.; Morgan, M.R.A.; Povey, M.J.W.; Day, A.J. Particle-stabilizing effects of flavonoids at the oil-water interface. *J. Agric. Food Chem.* **2011**, *59*, 2636–2645. [CrossRef]
48. Vega, C.; Grover, M.K. Physicochemical properties of acidified skim milk gels containing cocoa flavanols. *J. Agric. Food Chem.* **2011**, *59*, 6740–6747. [CrossRef]
49. Güçlü-Üstündağ, Ö.; Mazza, G. Saponins: Properties, Applications and Processing. *Crit. Rev. Food Sci. Nutr.* **2007**, *47*, 231–258. [CrossRef] [PubMed]
50. Randall, R.C.; Phillips, G.O.; Williams, P.A. The role of the proteinaceous component on the emulsifying properties of gum arabic. *Food Hydrocoll.* **1988**, *2*, 131–140. [CrossRef]
51. Chung, C.; Sher, A.; Rousset, P.; Decker, E.A.; McClements, D.J. Formulation of food emulsions using natural emulsifiers: Utilization of quillaja saponin and soy lecithin to fabricate liquid coffee whiteners. *J. Food Eng.* **2017**, *209*, 1–11. [CrossRef]

MDPI

St. Alban-Anlage 66

4052 Basel

Switzerland

Tel. +41 61 683 77 34

Fax +41 61 302 89 18

www.mdpi.com

Foods Editorial Office

E-mail: foods@mdpi.com

www.mdpi.com/journal/foods